T0297910

SPACE
TIME AND MATTER

SPACE TIME AND MATTER

Dipak K Sen

University of Toronto, Canada

World Scientific

NEW JERSEY • LONDON • SINGAPORE • BEIJING • SHANGHAI • HONG KONG • TAIPEI • CHENNAI

Published by

World Scientific Publishing Co. Pte. Ltd.
5 Toh Tuck Link, Singapore 596224
USA office: 27 Warren Street, Suite 401-402, Hackensack, NJ 07601
UK office: 57 Shelton Street, Covent Garden, London WC2H 9HE

Library of Congress Cataloging-in-Publication Data
Sen, D. K. (Dipak Kumar), 1934- author.
Space, time and matter / by Dipak K. Sen, University of Toronto, Canada.
pages cm
Includes bibliographical references and index.
ISBN 978-9814522830 (hardcover : alk. paper)
1. Space and time--Mathematical models. 2. Matter--Mathematical models. 3. Kinematics.
I. Title.
QC173.59.S65S466 2014
530.11--dc23

2014006245

British Library Cataloguing-in-Publication Data
A catalogue record for this book is available from the British Library.

Typeset by Stallion Press
Email: enquiries@stallionpress.com

Printed in Singapore

PREFACE

In 1918, Hermann Weyl wrote a book entitled *Raum-Zeit-Materie.* This short monograph, with the same title, however, treats the same subject matter in a somewhat unconventional way.

We present a novel, albeit equivalent, formalism of relativistic kinematics and general relativistic field dynamics in which time does not have the same primary role as space as in conventional relativity.

Finally, we present a theory of formation of fundamental particles where the fundamental constituents are left-handed and right-handed 2-component Weyl neutrinos.

The material presented here uses the technics of geometry of manifolds extensively, in particular, that of vector fields on manifolds. Readers unfamiliar with this subject should start first with Appendices A and B.

Dipak K. Sen
Toronto, Canada
December, 2013

CONTENTS

1. SPACE AND TIME

1.0. Introduction

It was Minkowski who paraphrased that, "...Henceforth space by itself, and time by itself, are doomed to fade away into mere shadows, and only a kind of union of the two will preserve an independent reality." And ever since relativists are accustomed to always think in the framework of a 4-dimensional space-time, it has almost become second nature to them.

Strictly speaking, however, the Minkowski signature $(- + + +)$ makes a clear distinction as to which coordinates can be interpreted as spatial and which ones as temporal, and recently, the usefulness of $3 + 1$ decomposition techniques has shown that sometimes the *separation* of space from time provides a better insight to what is going on rather than the traditional *fusion* of space with time.

Imagine a universe in which nothing moves. In such a universe, the notion of time disappears. So, space and motion in space are really the primary concepts and time is a secondary concept. It should thus be possible to construct a theory of Kinematic Relativity based on the primary notions of space and motion instead of space and time.

In this monograph, we present a new formulation of both relativistic kinematics and field dynamics in an entirely 3-dimensional space. The basic idea is to represent local physical observers as non-singular 3-dimensional local vector fields and the dynamics of physical fields by the "flows" of the vector fields, which are 1-parameter groups of local smooth transformations of the space. The flow parameter thus plays the role of local time for each physical observer, and Lie-derivative replaces the time derivative.

In the kinematics part, we first introduce the basic notions and postulates of our formalism which take place in a 3-dimensional Riemannian space (M^3, g). A physical observer is defined to be a non-singular vector field X on M^3 with $0 < g(X,X) < 1$. With the help of the metric, we can define a relative velocity function $V(X,Y)$ between any two physical observers X, Y. Two observers are then inertially equivalent if $V(X,Y) = \text{const.}$ on M^3. If $c : I \to M^3$ is a particle path, i.e., a smooth curve on M^3, we can define space and time intervals of $c(I)$ relative to a physical observer X in such a way that if Y is another physical observer inertially equivalent to X, then the space and time intervals relative to Y are related to that of X by a Lorentz transformation. The Lorentz time dilation formula is then an immediate consequence.

We next consider the problem of constructing an equivalence class of inertial observers starting from a representative physical observer. The problem leads us to the definition of "*generalized Lorentz matrices*" with corresponding "generalized" properties, which would reduce to the usual Lorentz matrices when (M^3, g) is an Euclidean space.

In the dynamics part, we first show how one can describe the time evolution of certain classical fields by subjecting the 3-dimensional fields to a transformation by the flow of a fundamental 3-dimensional vector field. We illustrate this with some simple 1-dimensional cases such as the 1-dimensional Heat and Wave equations. We next apply our basic procedure to the Gauss–Einstein equations, that is, the Einstein field equations in Gaussian normal coordinates, and obtain a set of equations for the 3-metric g and a 3-dimensional vector field X. Every solution (g, X) of these equations determine uniquely a space-time 4-metric solution of the Einstein field equations.

Our first example is the de Sitter solution. A generalization of the de Sitter case leads us to a *complex* metric of Kasner type, which in turn, provides us with a *new* (*real*) *solution* of the vacuum Einstein field equations, containing five real parameters. We then give another solution of our 3-dimensional equations which is mathematically interesting and non-trivial, but however, corresponds to the physically trivial flat space-time metric. Nevertheless, this solution clearly illustrates the basic principles of our 3-dimensional approach.

Finally, we apply our formalism to the Maxwell equations and obtain a corresponding 3-dimensional set of equations, which turn out to be, not only concise, but also elegant.

1.1. The Hyperbolic Structure of the Space of Relative Velocities

We first consider the geometry of the space of relative velocities in conventional Special Relativity. This will suggest the formalism of a theory of Kinematic Relativity on 3-manifolds which reproduces the essential features of conventional relativity.

In the special theory of relativity, the velocity transformation formula relating the relative velocities of three equivalent inertial systems $\mathbf{u}$, $\mathbf{v}$, $\mathbf{w}$ is given by (in units $c = 1$)

$$\|\mathbf{w}\|^{\mathbf{2}} = \frac{(\mathbf{u}-\mathbf{v})^{\mathbf{2}} - (\mathbf{u}\times\mathbf{v})^{\mathbf{2}}}{(\mathbf{1}-\mathbf{u}\cdot\mathbf{v})^{\mathbf{2}}}. \tag{1.1}$$

Equation (1.1) has the following geometrical significance. Let us choose some *reference* inertial system and consider the velocities $u = (u^1, u^2, u^3)$ of all other equivalent inertial systems relative to this reference inertial system. Since for *physical* inertial systems $\|u\| = [|\sum(u^i)^2|]^{1/2} < 1$, the space of relative velocities is *topologically* a 3-dimensional open disk

$$D^3 = \{u \in \mathbf{R}^3 | \|u\| < 1\}.$$

Note: For *tachyons*, the appropriate space to consider would be the closure of the complement of D^3, that is, $\mathbf{R}^3 \backslash \bar{D^3}$.

Now, we introduce the standard (positive definite) hyperbolic metric G on D^3 by

$$\begin{aligned} ds^2 &= \sum_{i,k} G_{ik}(u) du^i du^k \\ &= \frac{[1-\sum(u^i)^2][\sum(du^i)^2] + [\sum u^i du^i]^2}{[1-\sum(u^i)^2]^2}, \end{aligned} \tag{1.2}$$

that is,

$$G_{ik}(u) = \frac{[1 - \sum (u^i)^2]\delta_{ik} + u^i u^k}{[1 - \sum (u^i)^2]^2}.$$

Then, $\mathbf{H}^3 = (D^3, G)$ is the standard hyperbolic 3-space of constant curvature -1. This is the so-called Poincaré disk model of $\mathbf{H}^3$ ([1]). Alternatively, one can introduce homogeneous coordinates $\xi = (\xi_0, \xi_1, \xi_2, \xi_3)$ by $\xi_i = \xi_0 u^i$ and define for $\xi = (\xi_0, \xi_1, \xi_2, \xi_3), \eta = (\eta_0, \eta_1, \eta_2, \eta_3)$, the inner product $(\xi, \eta) = \xi_0 \eta_0 - \sum_i \xi_i \eta_i$. Then,

$$ds^2 = \frac{(d\xi, \xi)(\xi, d\xi) - (\xi, \xi)(d\xi, d\xi)}{(\xi, \xi)^2}. \tag{1.3}$$

$\mathbf{H^3}$ is geodesically complete and has a distance function d given by

$$d(u, v) = \cosh^{-1} \left[\frac{\left[1 - \sum_i u^i v^i\right]}{\left[1 - \sum_i (u)^2\right]^{1/2} \left[1 - \sum_i (v)^2\right]^{1/2}} \right] \tag{1.4}$$

or, in homogeneous coordinates,

$$d(\xi, \eta) = \cosh^{-1} \left[\frac{|(\xi, \eta)|}{(\xi, \xi)^{1/2} (\eta, \eta)^{1/2}} \right]. \tag{1.5}$$

A comparison of Eq. (1.1) with Eq. (1.5) suggests that the distance $d(u, v)$ is somehow related to the relative velocity w between two inertial systems whose velocities relative to the reference inertial system are u and v. To see the actual relationship, we should note that the reference system is at rest relative to itself. Now, from (1.4)

$$d(0, v) = \cosh^{-1} \left[\frac{1}{(1 - \|v\|^2)^{1/2}} \right] = \tanh^{-1} \|v\|.$$

That is, $\|v\| = \tanh d(0, v)$. We can, therefore, reinterpret Eq. (1.1) as a relation between the relative velocity w and the hyperbolic distance

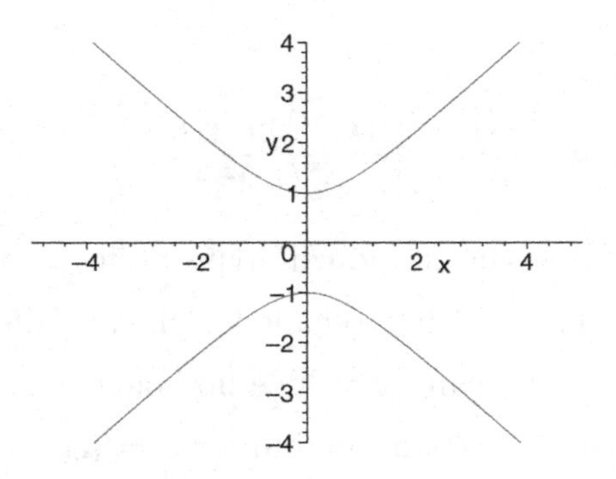

Fig. 1. The hypersurfaces $\mathbf{M}', \mathbf{M}''$.

$d(u,v)$ by

$$\|w\| = \tanh[d(u,v)]. \tag{1.6}$$

The isometry group of $\mathbf{H^3}$ is intimately connected with the Lorentz group, as can be seen in what follows.

First, we consider an *abstract* $\mathbf{R^4}$ with the *Lorentz* metric $g_{\mathbf{R^4}}$ given by $ds^2 = (dx^1)^2+(dx^2)^2+(dx^3)^2-(dx^4)^2$. Consider now the hypersurface M of $\mathbf{R^4}$: $M = \{x = (x^1,x^2,x^3,x^4) \in \mathbf{R}^4 | (x^1)^2+(x^2)^2+(x^3)^2-(x^4)^2 = -1\}$. M has two components: $\mathbf{M}' = \{x \in \mathbf{M} | x^4 \geq 1\}$ and $\mathbf{M}'' = \{x \in \mathbf{M} | x^4 \leq -1\}$ (see Fig. 1; here (x,y) represents (x^1,x^4)).

Both $\mathbf{M}'$ and $\mathbf{M}''$ are *diffeomorphic* to D^3 and the diffeomorphism map is given by

$$(x^1,x^2,x^3,x^4) \mapsto (u^1,u^2,u^3),$$
$$u^i = x^i/x^4, \quad (i=1,2,3)$$

with its inverse

$$(u^1,u^2,u^3) \mapsto (x^1,x^2,x^3,x^4),$$
$$x^i = u^i x^4, \quad x^4 = \pm\left[\frac{1}{(1-\sum(u^i)^2)}\right]^{1/2}.$$

Now, let g_M be the induced *Riemannian* metric on M obtained from $g_{\mathbf{R^4}}$ on $\mathbf{R^4}$. In coordinates (u^i), g_M is given exactly by (1.2). Thus, $\mathbf{H}^3 = (D^3, G)$ is *isometric* to $(M', g_{M|M'})$ and $(M'', g_{M|M''})$.

The isometry group of $\mathbf{H}^3$ was shown by Poincaré to be $PSL(2, \mathbf{C})$, the projective special linear group in two complex dimensions and is, thus, isomorphic to $L_+^\uparrow$, the *orthochronous* Lorentz group. It is clear from (1.5) that the isometry group of $\mathbf{H}^3$ leaves the inner product (ξ, η), in homogeneous coordinates, invariant. Let its action be given by

$$(\xi_0, \xi_1, \xi_2, \xi_3) \mapsto (\overline{\xi}_0, \overline{\xi}_1, \overline{\xi}_2, \overline{\xi}_3),$$

where

$$\overline{\xi}_\mu = L^\mu_\nu \xi_\nu, \quad L^\mu_\nu \in L_+^\uparrow \quad (\mu, \nu = 0, 1, 2, 3).$$

Since $\overline{u}^i = \overline{\xi}_i \overline{\xi}_0$, $L_+^\uparrow$ acts on $\mathbf{H}^3$ as a fractional linear transformation:

$$u^i \mapsto \overline{u}^i = (L^i_0 + L^i_k u^k)/(L^0_0 + L^0_k u^k).$$

Thus, the fundamental aspect of the Special Theory of Relativity, namely, Lorentz invariance, is contained in the hyperbolic structure of the space of relative velocities.

1.2. Relativistic Kinematics on 3-Manifolds

Let (M^3, g) be a 3-dimensional space with a (positive-definite) Riemannian metric g. By an observer (or a particle path), we usually mean a smooth curve in M^3, i.e., a smooth map $c : I \to M^3$ from a real interval I into M^3.

Since, to every such curve c, one can associate a local vector field C such that c is an integral curve of C ([2]). *We shall regard the set of all non-singular vector fields on M^3 as the set of all observers in M^3.*

Non-singularity is necessary to avoid the possibility that an observer comes to an absolute rest relative to M^3. To each vector field X corresponds a local flow χ_t, i.e., a local 1-parameter group of local transformations of M^3. Here, the parameter t measures the "flow of time" as perceived by the observer represented by X. In this sense, time is a local concept since X need not be globally defined and is not necessarily *complete* unless M^3 is compact ([3]).

Let us now consider only those non-singular vector fields X such that $g(X, X) < 1$. Such observers will be called physical observers.

(Corresponding to the fact that no such observer can attain the velocity of light.)

From now on, we consider only physical observers and denote by $\Sigma(M^3)$ the set of all physical observers.

We define a map V, called the *relative velocity map*, from a pair of physical observers to a smooth *function* on M^3 as follows:

$$V : \Sigma(M^3) \times \Sigma(M^3) \mapsto C^\infty(M^3),$$
$$(X, Y) \mapsto V(X, Y),$$

where

$$\begin{aligned} V(X,Y) &= \left[1 - \frac{1}{[F(X,Y)]^2}\right]^{1/2}, \\ F(X,Y) &= \frac{1 - g(X,Y)}{[1 - g(X,X)]^{1/2}[1 - g(Y,Y)]^{1/2}}. \end{aligned} \tag{1.7}$$

Here, $C^\infty(M^3)$ denotes the set of all smooth functions on M^3. Let us now define an equivalence relation $\sim$ in $\Sigma(M^3)$ by:

$$\begin{aligned} &X \sim Y \quad \text{if and only if} \\ &V(X,Y) = \text{constant function on } M^3. \end{aligned} \tag{1.8}$$

In other words, the observers X and Y are said to be (*inertially*) *equivalent* if the relative velocity function is constant on M^3.

We shall now give theoretical definitions of space and time intervals of a physical particle path relative to any physical observer X. Let $c : I \to M^3$, $\lambda \mapsto c(\lambda)$ be a particle path and C any representative corresponding vector field associated with it such that c is an integral curve of C. For a physical particle path, $0 < g(C,C) < 1$ on $c(I)$.

Definition. The *time interval of* c *relative to an observer* X, denoted by Δt_X, is given by the following integral over I:

$$\Delta t_X = \int_I F(X,C)d\lambda. \tag{1.9}$$

Note that we can reparametrize c by t_X, where $dt_X/d\lambda = F(X,C)$.

Definition. The *space interval of c relative to the observer X*, denoted by Δd_X, is given by

$$\Delta d_X = \int_I V(X,C)dt_X = \int_I V(X,C)\,F(X,C)d\lambda. \tag{1.10}$$

If $(\Delta t_Y, \Delta d_Y)$ are time and space intervals of c relative to another observer Y, then $(\Delta t_X, \Delta d_X) \to (\Delta t_Y, \Delta d_Y)$ provides a space-time transformation from observer X to observer Y.

1.3. Class of Inertial Observers and Generalized Lorentz Matrices

We now consider the problem of constructing a class of inertial observers starting from a representative physical observer. In other words, given a physical observer u, the problem of constructing an observer w such that u and w are inertially equivalent, that is, the relative velocity between u and w, $V(u,w)$, as given by (1.7), is a constant function on M^3.

Since M^3 is 3-dimensional, it is *parallelizable*, that is, there exists a global "framing" of M^3 by means of three linearly independent vector fields $\{e_{(i)}\}$, $i = 1,2,3$, so that any vector field u can be expressed as $u = u^i(x)e_{(i)}$, and $g(e_{(i)}, e_{(j)}) = g_{ij}(x)$, where $u^i(x)$, $g_{ij}(x)$ are functions on M^3. One can, of course, also consider everything locally in some local coordinate system (x^i), so that $u^i(x)$, $g_{ij}(x)$ are functions of the coordinate (x^i). Since u is a physical observer, we have $g(u,u) = g_{ik}u^i u^k = u_i u^i < 1$, where $u_i = g_{ik}u^k$.

Let

$$\left.\begin{aligned} \gamma_u &= [1 - g(u,u)]^{-1/2} \\ A_u &= (\gamma_u - 1)/g(u,u) \end{aligned}\right\} \tag{1.11}$$

and let us define the 4×4 matrix functions $\tilde{L}^{\mu}_{\lambda}(u)$, $\mu, \lambda = 0,1,2,3$ as follows[a]:

$$\left.\begin{aligned} \tilde{L}^0_0(u) &= \gamma_u, & \tilde{L}^0_k(u) &= \gamma_u u_k \\ \tilde{L}^k_0(u) &= \gamma_u u^k, & \tilde{L}^k_j(u) &= \delta^k_j + A_u u^k u_j. \end{aligned}\right\} \tag{1.12}$$

Note that in (1.8) u_j differ from u^k by the lowering of indices by g_{jk} and also that $\tilde{L}^{\mu}_{\lambda}$ are *functions* on M. These "generalized Lorentz

[a]In what follows Latin indices $i, j, k, \ldots = 1,2,3$ and the Greek indices $\alpha, \beta, \gamma, \ldots = 0,1,2,3$.

matrices" $\tilde{L}^{\mu}_{\lambda}(u)$ are easily seen to satisfy the following "generalized" properties:

$$\left.\begin{aligned} g_{ik}\tilde{L}^{i}_{0}\tilde{L}^{k}_{0} &= (\tilde{L}^{0}_{0})^2 - 1 \\ g_{mn}\tilde{L}^{m}_{0}\tilde{L}^{n}_{i} &= \tilde{L}^{0}_{0}\tilde{L}^{0}_{i} \\ g_{mn}\tilde{L}^{m}_{i}\tilde{L}^{n}_{j} &= \tilde{L}^{0}_{i}\tilde{L}^{0}_{j} + g_{ij} \end{aligned}\right\} \tag{1.13}$$

and when $(M^3, g) = (\mathbb{R}^3, \delta)$, that is, when (M^3, g) is the Euclidean space $\mathbb{R}^3$ with the Euclidean metric $g_{ik} = \delta_{ik}$, $\tilde{L}^{\mu}_{\lambda}(u)$ reduces to the usual Lorentz matrix $L^{\mu}_{\lambda}(\mathbf{u})$ with the pure boost given by $\mathbf{u} = (u^1, u^2, u^3)$ and satisfy the usual properties:

$$\left.\begin{aligned} L^{i}_{0}L^{i}_{0} &= (L^{0}_{0})^2 - 1 \\ L^{m}_{0}L^{m}_{i} &= L^{0}_{0}L^{0}_{i} \\ L^{m}_{i}L^{m}_{j} &= L^{0}_{i}L^{0}_{j} + \delta_{ij}. \end{aligned}\right\} \tag{1.14}$$

Let $v = v^i(x)e_{(i)}$ be another vector field where $g(v, v) < 1$.

We now define a vector field $w = w^i(x)e_{(i)}$ by

$$w^i = \frac{\tilde{L}^{i}_{0}(u) + \tilde{L}^{i}_{k}(u)v^k}{\tilde{L}^{0}_{0}(u) + \tilde{L}^{0}_{k}(u)v^k}. \tag{1.15}$$

Then it follows that $g(w, w) < 1$, so that w also represents a physical observer.

Using the "generalized" properties (1.9) one now shows that the relative velocity function $V(u, w)$ as given by (1.1) satisfies:

$$[V(u, w)]^2 = g(v, v).$$

So, if v is chosen so that $g(v, v) = \text{const.}$, the observers u and w are inertially equivalent and, in this way, starting from u, we can construct a class of inertially equivalent observers.

1.4. Classical Relativistic Field Dynamics

In classical field theory, the dynamics of a field is usually described by a set of evolution equations (together possibly with some constraint equations) in some space-time coordinates. We shall now demonstrate that in many cases the dynamics can also be described in a purely spatial (i.e., 3-dimensional) setting without the necessity of introducing explicitly a time-coordinate. Instead of the time-coordinate and time-dependent quantities, the temporal evolution is provided by the "flow" of a 3-dimensional vector field representing a physical observer.

Before considering the Einstein field equations in some detail, we shall first illustrate our basic approach with some simple examples from classical field theory, such as the Heat and Wave equations. For simplicity, we restrict our analysis to the 1-dimensional case only. The generalization to three spatial dimensions is obvious.

1.4.1. *One-dimensional Heat equation*

Consider the 1-dimensional Heat equation with some initial data:

$$\left.\begin{aligned} \psi_t &= \psi_{xx} \\ \psi(x,0) &= \phi(x). \end{aligned}\right\} \tag{1.16}$$

We look for solutions $\psi(x,t)$ of (1.16) which come from the initial data $\phi(x)$ through a transformation by the flow of a vector field X depending on the spatial coordinate only, as follows. Let X be given by

$$X = \alpha(x)\frac{d}{dx}. \tag{1.17}$$

Consider its integral curves given by the dynamical system

$$\frac{dx}{dt} = \alpha(x). \tag{1.18}$$

That is, $\int dt = \int \frac{dx}{\alpha(x)} = \nu(x)$, say. So that $x(t) = \nu^{-1}(t+c)$, $c =$ const., and $x(0) = \nu^{-1}(c)$ or $c = \nu(x(0))$. The flow χ_t of X transforms the initial point $x(0)$ onto the point $x(t)$, that is,

$$\chi_t : x \mapsto \chi_t(x) = \nu^{-1}(t + \nu(x)). \tag{1.19}$$

Now let[b]

$$\left.\begin{aligned} \psi(x,t) &= (\tilde{\chi}_t\phi)(x) \\ &= \phi(\chi_t(x)) \\ &= \phi(\nu^{-1}(t+\nu(x))) \end{aligned}\right\} \tag{1.20}$$

so that

$$\psi(x,0) = \phi(x).$$

Then $\psi(x,t)$, which now also depends on the flow parameter t, is the transformed initial data $\phi(x)$, which of course depends only on the space variable x. Let us now substitute $\psi(x,t)$ in (1.16).

Put $t + \nu(x) = z$, and $\nu(x) = y$. Then, $\nu^{-1\prime}(y) = 1/\nu'(x) = \alpha(x)$; $\nu^{-1\prime\prime}(y) = \alpha'(x)\alpha(x)$. And

$$\left.\begin{aligned} \psi_t(x,t) &= \phi'\big(\nu^{-1}(z)\big)\nu^{-1\prime}(z) \\ \psi_t(x,0) &= \phi'(x)\alpha(x) = \mathcal{L}_X\phi \end{aligned}\right\} \tag{1.21}$$

$$\left.\begin{aligned} \psi_x(x,t) &= \phi'\big(\nu^{-1}(z)\big)\nu^{-1\prime}(z)\nu'(x) \\ \psi_x(x,0) &= \phi'(x)\nu^{-1\prime}(y)\nu'(x) = \phi'(x) \end{aligned}\right\} \tag{1.22}$$

---

[b]Geometrical objects (e.g., scalar, vector, tensor fields and densities) can be transformed by the flow of a vector field ([3]). We could have also transformed $\phi(x)$ by ${\chi_t}^{-1} = \chi_{-t}$, by reversing the flow, that is, by letting $\psi(x,t) = \phi({\chi_t}^{-1}(x))$. This would have preserved the scalar character of ϕ under the flow.

$$\psi_{xx}(x,t) = \phi''(\nu^{-1}(z))\left[\nu^{-1\prime}(z)\right]^2\left[\nu'(x)\right]^2 + \phi'(\nu^{-1}(z))\nu^{-1\prime\prime}(z)[\nu'(x)]^2 + \phi'(\nu^{-1}(z))\nu^{-1\prime}(z)\nu''(x). \tag{1.23}$$

$$\begin{aligned}\psi_{xx}(x,0) &= \phi''(x)\left[\nu^{-1\prime}(y)\right]^2\left[\nu'(x)\right]^2 + \phi'(x)\nu^{-1\prime\prime}(y)\left[\nu'(x)\right]^2 \\ &\quad + \phi'(x)\nu^{-1\prime}(y)\nu''(x) \\ &= \phi''(x)\left[\alpha(x)\right]^2\left[\alpha(x)\right]^{-2} \\ &\quad + \phi'(x)\alpha'(x)\alpha(x)\left[\alpha(x)\right]^{-2} \\ &\quad - \phi'(x)\alpha(x)\alpha'(x)\left[\alpha(x)\right]^{-2} = \phi''(x).\end{aligned} \tag{1.24}$$

If our $\psi(x,t)$ is to satisfy (1.16), we must have

$$\begin{aligned}\phi'(\nu^{-1}(z))\nu^{-1\prime}(z) &= \phi''(\nu^{-1}(z))\left[\nu^{-1\prime}(z)\right]^2\left[\nu'(x)\right]^2 \\ &\quad + \phi'(\nu^{-1}(z))\nu^{-1\prime\prime}(z)\left[\nu'(x)\right]^2 \\ &\quad + \phi'(\nu^{-1}(z))\nu^{-1\prime}(z)\nu''(x).\end{aligned} \tag{1.25}$$

Let us put $\nu^{-1}(z) = u$. So that again, $\nu^{-1\prime}(z) = \frac{1}{\nu'(u)} = \alpha(u)$, $\nu^{-1\prime\prime}(z) = \alpha'(u)\alpha(u)$. We can write (1.25) as

$$\begin{aligned}\phi'(u)\alpha(u) &= \left[\phi''(u)\left[\alpha(u)\right]^2 + \phi'(u)\alpha'(u)\alpha(u)\right]\left[\nu'(x)\right]^2 \\ &\quad + \phi'(u)\alpha(u)\nu''(x).\end{aligned}$$

The variables x and u can thus be separated to give

$$\frac{1-\nu''(x)}{[\nu'(x)]^2} = \frac{\phi''(u)}{\phi'(u)}\alpha(u) + \alpha'(u)$$

or

$$\left[\alpha(x)\right]^2 + \alpha'(x) = \frac{\phi''(u)}{\phi'(u)}\alpha(u) + \alpha'(u) = \kappa(\text{const.}) \tag{1.26}$$

which must be satisfied for all x and u. In particular, for $u = x$, we must have from (1.26)

$$\left.\begin{aligned} \alpha(x)\phi'(x) &= \phi''(x) \\ \text{or} \qquad \mathcal{L}_X\phi &= \Delta\phi. \end{aligned}\right\} \tag{1.27}$$

Note that (1.27) also follows directly from (1.16) and the initial values in (1.21), (1.24), and that (1.27) can be obtained from (1.16) by *replacing* $\psi(x,t)$ *with* $\phi(x)$ *and the time-derivative* $\frac{\partial}{\partial t}$ *with the Lie-derivative* $\mathcal{L}_X$ *with respect to* X. Any solution (ϕ, X) of (1.24) provides a solution ψ of (1.16), because from (1.27), ([3]),

$$\frac{\partial}{\partial t}\psi = \frac{\partial}{\partial t}(\tilde{\chi}_t\phi) = \tilde{\chi}_t(\mathcal{L}_X\phi) = \tilde{\chi}_t(\Delta\phi) = \Delta(\tilde{\chi}_t\phi) = \Delta\psi. \tag{1.28}$$

Since (1.26) or (1.27) *are equations for both* X *and* ϕ, *the initial data cannot be arbitrary* and, therefore, our procedure provides only special solutions of (1.16). We shall give two examples of such solutions ([4]).

Example 1. $\phi(x) = e^{\alpha x}$, $\alpha(x) = \alpha = \text{const.}$, $X = \alpha\frac{d}{dx}$. Then (1.27) is satisfied, and $\nu(x) = \frac{x}{\alpha}$, $\nu^{-1}(y) = \alpha y$, $\nu^{-1}\big(t + \nu(x)\big) = \alpha t + x$. So $\psi(x,t) = \phi\big(\nu^{-1}(t + \nu(x))\big) = \phi(\alpha t + x) = e^{\alpha^2 t + \alpha x}$. From (1.26), we get $\alpha^2 = \kappa$. If we also allow κ to be negative, we also obtain complex solutions $\psi(x,t) = e^{\alpha^2 t \pm i\beta x}$ with $\alpha = \pm i\beta$.

Example 2. $\phi(x) = x^2$, $\alpha(x) = \frac{1}{x}$, $X = \frac{1}{x}\frac{d}{dx}$. Again, (1.27) is satisfied and $\nu(x) = \frac{x^2}{2}$, $\nu^{-1}(y) = \pm\sqrt{2y}$, $\nu^{-1}\big(t+\nu(x)\big) = \pm\sqrt{2t + x^2}$. So $\psi(x,t) = \phi\big(\nu^{-1}(t + \nu(x))\big) = \phi(\pm\sqrt{2t + x^2}) = 2t + x^2$. This is a well-known polynomial solution of (1.16).

In fact, it is easily possible to obtain the general solution of (1.26) or (1.27).

1.4.2. One-dimensional Wave equation

Consider now the 1-dimensional Wave equation with some initial data:

$$\left.\begin{aligned} \psi_{tt} &= \psi_{xx} \\ \psi(x,0) &= \phi(x) \\ \psi_t(x,0) &= \eta(x). \end{aligned}\right\} \tag{1.29}$$

Again let $X = \alpha(x)\frac{d}{dx}$ and its flow χ_t be given by (1.16), where $\nu'(x) = \frac{1}{\alpha(x)}$, and $\psi(x,t) = \phi(\nu^{-1}(t+\nu(x)))$. Now since $\psi_t(x,0) = \mathcal{L}_X\phi$, we must have

$$\mathcal{L}_X\phi = \eta. \tag{1.30}$$

From (1.24)

$$\left.\begin{aligned} \psi_{tt}(x,t) &= \phi''\big(\nu^{-1}(z)\big)\big[\nu^{-1\prime}(z)\big]^2 + \phi'\big(\nu^{-1}(z)\big)\nu^{-1\prime\prime}(z) \\ \psi_{tt}(x,0) &= \phi''(x)\big[\alpha(x)\big]^2 + \phi'(x)\alpha'(x)\alpha(x). \end{aligned}\right\} \tag{1.31}$$

It follows from (1.24), (1.29), (1.31) that

$$\left.\begin{aligned} &\phi''(x)\big[\alpha(x)\big]^2 + \phi'(x)\alpha'(x)\alpha(x) = \phi''(x) \\ \text{or}\quad &\mathcal{L}_X\mathcal{L}_X\phi = \Delta\phi. \end{aligned}\right\} \tag{1.32}$$

Again (1.32) is obtained from (1.29) by replacing $\psi(x,t)$ by $\phi(x)$ and the time-derivative $\frac{\partial}{\partial t}$ by the Lie-derivative $\mathcal{L}_X$. Any solution of (1.30), (1.32) provides a solution of (1.29). Note that *the initial data ϕ and η now have to be related by the "constraint equation"* (1.30) and (ϕ, X) must satisfy (1.32). We give two special solutions of (1.30) and (1.32), and hence of (1.29).

(i) $\alpha(x) = +1$, $X = \frac{d}{dx}$, $\nu(x) = x$, $\nu^{-1}\big(t+\nu(x)\big) = t + x$, $\psi(x,t) = \phi(x+t)$ with $\eta(x) = \phi'(x)$,
(ii) $\alpha(x) = -1$, $X = -\frac{d}{dx}$, $\nu(x) = -x$, $\nu^{-1}\big(t+\nu(x)\big) = -t + x$, $\psi(x,t) = \phi(x-t)$ with $\eta(x) = -\phi'(x)$.

It is also easily possible to obtain the general solution of (1.30) and (1.29).

1.4.3. *Gauss–Einstein equations*

In conventional General Relativity the space-time metric $g_{\alpha\beta}$ satisfies the Einstein field equations, which are basically evolution equations with certain constraints, and it is well known ([5]) that the *vacuum* Einstein field equations: $R_{\alpha\beta} = 0$ *equivalent* to

$$\left.\begin{aligned} G^0_\alpha &= 0 \quad (\textit{constraints}\ \text{equations}) \\ R_{ik} &= 0 \quad (\textit{evolution}\ \text{equations}) \end{aligned}\right\} \tag{1.33}$$

provided that $g^{00} \neq 0$; for example, on a non-null hypersurface. Here $G_{\alpha\beta} \equiv R_{\alpha\beta} - \frac{1}{2}g_{\alpha\beta}R$ is the Einstein tensor. A slight modification is necessary if one wishes to include a non-zero cosmological constant Λ in the field equations. The vacuum Einstein field equations with a cosmological constant: $R_{\alpha\beta} = \Lambda g_{\alpha\beta}$ also split up into a set of constraint and a set of evolution equations and they take a particularly simple form if one uses a *Gaussian normal coordinate* system in which $g_{00} = g^{00} = -1$ and $g_{0i} = g^{0i} = 0$. Then the field equations:

$$\left.\begin{aligned} R_{0i} &= 0 \\ R_{00} &= -\Lambda \\ R_{ik} &= \Lambda g_{ik} \end{aligned}\right\} \tag{1.34}$$

are *equivalent* to:

$$\left.\begin{aligned} &\left.\begin{aligned} G^0_i &= 0 \\ G^0_0 &= -\Lambda \end{aligned}\right\} \quad (\textit{constraint}\ \text{equations}) \\ &R_{ik} = \Lambda g_{ik} \quad (\textit{evolution}\ \text{equations}) \end{aligned}\right\} \tag{1.35}$$

because (1.34) *implies* (1.35), since $G_i^0 = g^{00}R_{i0} + g^{0k}R_{ik} = 0$ and $G_0^0 = \frac{1}{2}g^{00}R_{00} - \frac{1}{2}g^{ik}R_{ik} = \frac{1}{2}\Lambda(g^{00}g_{00} - g^{ik}g_{ik}) = -\Lambda$.

Conversely, (1.35) *implies* (1.34), since if $G_i^0 = g^{00}R_{i0} + g^{0k}R_{ik} = 0$, then from (1.35) $R_{i0} = 0$; and if $G_0^0 = \frac{1}{2}g^{00}R_{00} - \frac{1}{2}g^{ik}R_{ik} = -\Lambda$, then from (1.35), $\frac{1}{2}g^{00}R_{00} - \frac{3}{2}\Lambda = -\Lambda$, that is, $R_{00} = -\Lambda$.

The corresponding *matter* field equations with an energy momentum tensor $T_{\alpha\beta} : G_{\alpha\beta} + \Lambda g_{\alpha\beta} + T_{\alpha\beta} = 0$ or, equivalently, $R_{\alpha\beta} = \frac{1}{2}Tg_{\alpha\beta} - T_{\alpha\beta} + \Lambda g_{\alpha\beta}$ take the following form (in the case of *dustlike* matter with density ρ, pressure $p = 0$, and also in a *Gaussian normal coordinate* system):

$$\left.\begin{aligned} R_{0i} &= 0 \\ R_{00} &= -\frac{1}{2}\rho - \Lambda \\ R_{ik} &= \left(-\frac{1}{2}\rho + \Lambda\right) g_{ik}. \end{aligned}\right\} \tag{1.36}$$

Equation (1.36) is then equivalent to

$$\left.\begin{cases} \left.\begin{aligned} G_i^0 &= 0 \\ G_0^0 &= \rho - \Lambda \end{aligned}\right\} & (\textit{constraint} \text{ equations}) \\ R_{ik} = \left(-\frac{1}{2}\rho + \Lambda\right) g_{ik} & (\textit{evolution} \text{ equations}). \end{cases}\right\} \tag{1.37}$$

Gaussian normal coordinates can be introduced in general always, and with $x = (x^0 = t, x^1, x^2, x^3)$, the metric takes the form: $ds^2 = -dt^2 + g_{ik}(x)dx^i dx^k$. So that $g_{00} = g^{00} = -1$, $g_{0i} = g^{0i} = 0$. The 4-metric $g_{\alpha\beta}$ thus determines a 3-metric g_{ik}, with its inverse g^{ik}, which is positive definite on the hypersurfaces $t =$ const. The field equations (1.34) are then explicitly, in a Gaussian normal coordinate system ([6]):

$$G_i^0 \equiv \frac{1}{2}(g_{i\ell,0})_{;\ell} - \frac{1}{2}(g^{\ell k} g_{\ell k,0})_{,i} = 0, \tag{1.38}$$

$$G_0^0 \equiv -\frac{1}{2}\overline{R} + \frac{1}{8} g^{\ell m} g_{\ell m,0} g^{ik} g_{ik,0}$$
$$- \frac{1}{8} g^{i\ell} g^{km} g_{ik,0} g_{\ell m,0} = \rho - \Lambda, \tag{1.39}$$

$$R_{ik} \equiv \overline{R}_{ik} - \frac{1}{2} g_{ik,00} - \frac{1}{4} g^{\ell m} g_{\ell m,0} g_{ik,0} + \frac{1}{2} g^{\ell m} g_{i\ell,0} g_{km,0}$$
$$= \left(-\frac{1}{2}\rho + \Lambda\right) g_{ik} \tag{1.40}$$

where

$$g_{ik,0} \equiv \partial g_{ik}/\partial x^0 = \partial g_{ik}/\partial t,$$
$$g_{ik,00} \equiv \partial^2 g_{ik}/\partial t^2,$$
$$\overline{R}_{ik} \equiv \text{Ricci-tensor of the 3-metric } g_{ik},$$
$$\overline{R} \equiv \text{Ricci-scalar of the 3-metric } g_{ik},$$
$$A_{;\ell} \equiv \text{covariant derivative of } A.$$

We shall refer to Eqs. (1.38)–(1.40) as the *Gauss–Einstein* field equations.[c]

[c]Solutions of these equations have been considered by Marsden and Fischer ([7]) in an infinite dimensional setting. Our approach to the problem is, however, entirely finite dimensional.

1.5. Relativistic Field Dynamics on 3-Manifolds

1.5.1. *Three-dimensional field equations and relationship with Einstein equations*

As we have seen that in classical field theory, the dynamics of a field is usually described by a set of evolution equations (together possibly with some constraint equations) in some space-time manifold. We shall now demonstrate that in many cases the dynamics can also be described in a purely 3-dimensional setting without the necessity of explicitly introducing a time-coordinate. Instead of the time-coordinate and time-dependent quantities, the temporal evolution is provided by the "flow" of a 3-dimensional vector field representing a fundamental observer.

The starting point of our dynamical theory would be a set of differential equations for a 3-metric g and a fundamental 3-vector field X on a M^3. Let $X = X^i(x)\partial/\partial x^i$ be a vector field on M^3 (in some local coordinate system (x^i)), and $h = \mathcal{L}_X g$, the Lie-derivative of g with respect to X.

So that

$$h = h_{ik} dx^i \otimes dx^k \quad \text{where } h_{ik} = (\mathcal{L}_X g)_{ik}.$$

Let

$$\begin{aligned} f_{,i}, f_{;i} &\equiv \text{partial and covariant derivative of } f\text{, respectively,} \\ R_{ik} &\equiv \text{Ricci-tensor of the 3-metric } g_{ik}, \\ R &\equiv \text{Ricci-scalar of the 3-metric } g_{ik}. \end{aligned}$$

Then our basic equations are:

$$h^{\ell}_{i;\ell} - (g^{\ell k} h_{\ell k})_{,i} = 0, \tag{1.41}$$

$$-\frac{1}{2}R + \frac{1}{8}(g^{\ell m}h_{\ell m})(g^{ik}h_{ik})$$
$$-\frac{1}{8}(g^{i\ell}h_{ik})(g^{km}h_{\ell m}) = 0, \tag{1.42}$$
$$R_{ik} - \frac{1}{2}(\mathcal{L}_X h)_{ik} - \frac{1}{4}(g^{\ell m}h_{\ell m})h_{ik}$$
$$+\frac{1}{2}(g^{\ell m}h_{i\ell})h_{km} = 0. \tag{1.43}$$

We regard Eqs. (1.41)–(1.43) as *a set of differential equations for both the 3-metric g and the 3-dimensional vector field X on M^3*. Every solution (g, X) of (1.41)–(1.43) determines uniquely a solution of the vacuum Einstein field equations as follows:

The integral curves of X determine a flow φ_τ, which are local 1-parameter group of local diffeomorphisms ([5]) of M^3. Thus, for each value of the flow parameter τ, φ_τ defines a point transformation: $x \mapsto \tilde{x} = \varphi_\tau(x)$ with $\varphi_0(x) = x$. The flow transformed metric $\tilde{g} = \tilde{\varphi}_\tau(g)$ is given by $\tilde{g}_{ik}(\tau, \tilde{x}) = \frac{\partial x^\ell}{\partial \tilde{x}^i}\frac{\partial x^m}{\partial \tilde{x}^k} g_{\ell m}(x)$, and is thus τ-dependent.

Theorem 1. *Let $(g_{ik}; X^i)$ be a solution of (1.9)–(1.11) and $\tilde{g}_{ik}(\tau, \tilde{x})$ the τ-dependent metric transformed by the flow φ_τ of X. Then, the 4-dimensional metric,*

$$ds^2 = -d\tau^2 + \tilde{g}_{ik}(\tau, \tilde{x})d\tilde{x}^i d\tilde{x}^k \tag{1.44}$$

is a solution of the vacuum Einstein equations in a Gaussian normal coordinate system $(\tilde{x}^0 = \tau, \tilde{x}^i)$.

Proof. Let

$K \equiv$ any tensor field,

$\tilde{\varphi}_\tau(K) \equiv$ the flow transformed tensor field K,

$\mathcal{L}_X K \equiv$ the Lie-derivative of K with respect to X.

Then one has ([5])

$$\tilde{\varphi}_\tau(\mathcal{L}_X K) = \frac{\partial}{\partial \tau}(\tilde{\varphi}_\tau(K)). \tag{1.45}$$

Therefore, by applying $\tilde{\varphi}_\tau$ to Eqs. (1.41)–(1.43), one can replace the Lie-derivatives of g_{ik} in (1.41)–(1.43) by the derivatives of $\tilde{g}_{ik}(\tau, \tilde{x}^i)$ with respect to τ, and obtain

$$(\tilde{g}^{k\ell}\tilde{g}_{ik,0})_{;\ell} - (\tilde{g}^{\ell k}\tilde{g}_{\ell k,0})_{,i} = 0, \tag{1.46}$$

$$-\frac{1}{2}\widetilde{R} + \frac{1}{8}\tilde{g}^{\ell m}\tilde{g}_{\ell m,0}\tilde{g}^{ik}\tilde{g}_{ik,0} - \frac{1}{8}\tilde{g}^{i\ell}\tilde{g}^{km}\tilde{g}_{ik,0}\tilde{g}_{\ell m,0} = 0, \tag{1.47}$$

$$\widetilde{R}_{ik} - \frac{1}{2}\tilde{g}_{ik,00} - \frac{1}{4}\tilde{g}^{\ell m}\tilde{g}_{\ell m,0}\tilde{g}_{ik,0} + \frac{1}{2}\tilde{g}^{\ell m}\tilde{g}_{i\ell,0}\tilde{g}_{km,0} = 0. \tag{1.48}$$

(Here, $\widetilde{R}_{ik}, \widetilde{R}$ and the covariant derivatives are all relative to the 3-metric $\tilde{g}_{ik}(\tau, \tilde{x}^i)$, and $\tilde{g}_{ik,0}$, $\tilde{g}_{ik,00}$ the first and second partial derivatives of $\tilde{g}_{ik}$ with respect to τ.) These are precisely the vacuum Einstein field equations for the metric (1.44) in the Gaussian normal coordinate system $(\tilde{x}^0 = \tau, \tilde{x}^i)$ ([6]). □

Note that the signature of the 4-metric (1.44) would depend on the signature of the 3-metric $\tilde{g}_{ik}$. *Thus, τ may or may not be the physical time-coordinate.*

1.6. Flat-Space Solutions

We now consider some special solutions of (1.41)–(1.43). First, note that if, either $X = 0$ or X is a *Killing vector field* of g, then $h = \mathcal{L}_X g = 0$. So that (1.41) is identically satisfied. Furthermore, if $\rho = \Lambda = 0$, then (1.43) implies that $\overline{R}_{ik} = 0$ which in turn implies that $\overline{R}_{ijk\ell} = 0$ (since M^3 is 3-dimensional), that is, M^3 is flat with the metric $g = \delta$ (i.e., $g_{ik} = \delta_{ik}$ in some coordinate system). So, $\{X = 0, g = \delta\}$ and $\{X$ is Killing, $g = \delta\}$ are both solutions of (1.41)–(1.43) with $\rho = \Lambda = 0$. These are *trivial* flat-space solutions.

For non-trivial flat-space solutions, we put $g_{ik} = g^{ik} = \delta_{ik}$, so that $\overline{R}_{ik} = 0$, $\overline{R} = 0$, and

$$
\begin{aligned}
h_{ik} &= X^i_{,k} + X^k_{,i}, \\
(\mathcal{L}_X h)_{ik} &= h_{ik,\ell} X^\ell + h_{\ell i} X^\ell_{,k} + h_{\ell k} X^\ell_{,i}, \\
h_{i\ell;\ell} &= h_{i\ell,\ell}; \quad (g^{\ell k} h_{\ell k})_{,i} = \left(\sum_\ell h_{\ell\ell} \right)_{,i},
\end{aligned} \tag{1.49}
$$

$$
\begin{aligned}
\frac{1}{8}(g^{i\ell} h_{ik})(g^{km} g_{\ell m}) = \frac{1}{8}(h_{11}^2 + h_{22}^2 + h_{33}^2) \\
+ \frac{1}{4}(h_{12}^2 + h_{13}^2 + h_{23}^2).
\end{aligned}
$$

Therefore, (1.41)–(1.43) becomes (together with (1.49)):

$$
\begin{aligned}
&\frac{1}{2}(h_{i1,1} + h_{i2,2} + h_{i3,3}) \\
&\qquad - \frac{1}{2}(h_{11,i} + h_{22,i} + h_{33,i}) = 0,
\end{aligned} \tag{1.50}
$$

$$
\begin{aligned}
&\frac{1}{4}(h_{11}h_{22} + h_{11}h_{33} + h_{22}h_{33}) \\
&\qquad - \frac{1}{4}(h_{12}^2 + h_{13}^2 + h_{23}^2) = \rho - \Lambda,
\end{aligned} \tag{1.51}
$$

$$-\frac{1}{2}(h_{ik,\ell}X^\ell + h_{\ell i}X^\ell_{,k} + h_{\ell k}X^\ell_{,i}) + \frac{1}{2}(h_{i1}h_{k1} + h_{i2}h_{k2} + h_{i3}h_{k3})$$

$$-\frac{1}{4}(h_{11} + h_{22} + h_{33})h_{ik} = \left(-\frac{1}{2}\rho + \Lambda\right)\delta_{ik}. \quad (1.52)$$

In terms of X^i and its derivatives, Eqs. (1.50)–(1.52) are

$$\frac{1}{2}(X^i_{,\ell} + X^\ell_{,i})_{,\ell} - (X^\ell_{,\ell})_{,i} = 0, \quad (1.53)$$

$$\frac{1}{2}(X^i_{,i})^2 - \frac{1}{8}(X^i_{,m} + X^m_{,i})(X^m_{,i} + X^i_{,m}) = \rho - \Lambda, \quad (1.54)$$

$$-\frac{1}{2}(X^i_{,k} + X^k_{,i})_{,\ell}X^\ell - \frac{1}{2}(X^i_{,\ell} + X^\ell_{,i})X^\ell_{,k}$$

$$-\frac{1}{2}(X^\ell_{,k} + X^k_{,\ell})X^\ell_{,i}$$

$$-\frac{1}{2}X^\ell_{,\ell}(X^i_{,k} + X^k_{,i}) + \frac{1}{2}(X^i_{,m} + X^m_{,i})(X^m_{,k} + X^k_{,m})$$

$$= \left(-\frac{1}{2}\rho + \Lambda\right)\delta_{ik}. \quad (1.55)$$

In spite of the complexity of the above equations, it is possible to find some non-trivial solutions as we shall see next.

1.7. The de Sitter Solution

As a non-trivial solution of (1.53)–(1.55) or (1.49)–(1.52), let $\rho = 0$ and $X^i = -\lambda x^i$, where λ is a constant to be determined. Then by (1.49) $(\mathcal{L}_X g)_{ik} = h_{ik} = -2\lambda\delta_{ik}$ and

$$-\frac{1}{2}(\mathcal{L}_X h)_{ik} = -\frac{1}{2}[-\lambda h_{\ell i}x^{\ell}_{,k} - \lambda h_{\ell k}x^{\ell}_{,i}] = -2\lambda^2\delta_{ik},$$

$$-\frac{1}{4}(g^{\ell m}h_{\ell m})h_{ik} = -\frac{1}{4}(\delta^{\ell m}h_{\ell m})h_{ik} = -3\lambda^2\delta_{ik},$$

$$\frac{1}{2}(g^{\ell m}h_{i\ell})h_{km} = \frac{1}{2}(\delta^{\ell m}h_{i\ell})h_{km} = 2\lambda^2\delta_{ik}.$$

So (1.50) is identically satisfied and (1.51)–(1.52) or (1.42)–(1.43), become

$$3\lambda^2 = -\Lambda, \tag{1.56}$$

$$-2\lambda^2\delta_{ik} + 2\lambda^2\delta_{ik} - 3\lambda^2\delta_{ik} = \Lambda\delta_{ik}. \tag{1.57}$$

These equations are satisfied by taking $\lambda = \sqrt{\frac{-\Lambda}{3}}$. So that $\{g = \delta,$ $X = -\sqrt{\frac{-\Lambda}{3}}x^i\frac{\partial}{\partial x^i}\}$, $\rho = 0$ is a solution of (1.49)–(1.52), that is, a flat-space solution of (1.41)–(1.43).

The flow generated by the vector field $X = -\lambda x^i\frac{\partial}{\partial x^i}$ is simply $x^i \mapsto \tilde{x}^i = e^{-\lambda t}x^i$, so that $\frac{\partial x^i}{\partial \tilde{x}^k} = e^{\lambda t}\delta_{ik}$ and the transformed time-dependent 3-metric is therefore $\tilde{g}_{ik}(\tilde{x}, t) = e^{2\lambda t}\delta_{ik}$. The corresponding 4-metric (in coordinates $(t, \tilde{x}^i)$):

$$ds^2 = -dt^2 + \tilde{g}_{ik}d\tilde{x}^i dx^k = -dt^2 + e^{2\lambda t}\delta_{ik}d\tilde{x}^i d\tilde{x}^k$$

with $\lambda = \sqrt{\frac{-\Lambda}{3}}$ is the well-known de Sitter solution of the Gauss–Einstein equations (with $\rho = 0$).

1.8. A New Solution of the Vacuum Einstein Field Equations

As our next non-trivial solution of (1.49)–(1.52), we make the ansatz: $\rho = \Lambda = 0$ and $X^1 = -\lambda_1 x^1$, $X^2 = -\lambda_2 x^2$, $X^3 = -\lambda_3 x^3$, i.e., $X^i = -\lambda_i x^i$ (no summation), where λ_i's are constants to be determined. Then $h_{ik} = -(\lambda_i + \lambda_k)\delta_{ik}$ (no summation), i.e., $h_{11} = -2\lambda_1$, $h_{22} = -2\lambda_2$, $h_{33} = -2\lambda_3$, $h_{ik} = 0$ if $i \neq k$.

Again, (1.50) is identically satisfied, and (1.51) implies

$$\lambda_1\lambda_2 + \lambda_1\lambda_3 + \lambda_2\lambda_3 = 0. \tag{1.58}$$

Now $-\frac{1}{2}(\mathcal{L}_X h)_{ik} = -\frac{1}{2}(h_{\ell i}X^\ell_{,k} + h_{\ell k}X^\ell_{,i})$. If $i \neq k$, for example, $i = 1$, $k = 2$, $-\frac{1}{2}(\mathcal{L}_X h)_{12} = 0$, $\frac{1}{2}(h_{11}h_{21} + h_{12}h_{22} + h_{13}h_{23}) = 0$ and $-\frac{1}{4}(h_{11} + h_{22} + h_{33})h_{12} = 0$. So that (1.52) is identically satisfied if $i \neq k$. If $i = k$, for example, $i = 1$, $k = 1$, $-\frac{1}{2}(\mathcal{L}_X h)_{11} = -2\lambda_1^2$, $\frac{1}{2}(h_{11}h_{11} + h_{12}h_{12} + h_{13}h_{13}) = 2\lambda_1^2$ and $-\frac{1}{4}(h_{11} + h_{22} + h_{33})h_{11} = -\lambda_1(\lambda_1 + \lambda_2 + \lambda_3)$. So that (1.52) implies

$$\left.\begin{aligned} -\lambda_1(\lambda_1 + \lambda_2 + \lambda_3) = 0 \\ -\lambda_2(\lambda_1 + \lambda_2 + \lambda_3) = 0 \\ -\lambda_3(\lambda_1 + \lambda_2 + \lambda_3) = 0. \end{aligned}\right\}$$

So, we would have a solution of the form $X^i = -\lambda_i x^i$ (no summation) provided

$$\lambda_1 + \lambda_2 + \lambda_3 = 0$$

and

$$\lambda_1\lambda_2 + \lambda_1\lambda_3 + \lambda_2\lambda_3 = 0$$

which is equivalent to:

$$\left.\begin{array}{r}\lambda_1+\lambda_2+\lambda_3=0\\ \lambda_1^2+\lambda_2^2+\lambda_3^2=0\end{array}\right\} \tag{1.59}$$

which is possible only if, either all λ_i are zero or some of the λ_i are *complex*. Let us entertain the possibility that some of the λ_i are *complex*.

The flow generated by the vector field $X=-(\lambda_1 x^1\frac{\partial}{\partial x^1}+\lambda_2 x^2\frac{\partial}{\partial x^2}+\lambda_3 x^3\frac{\partial}{\partial x^3})$ is now $x^i\mapsto\tilde{x}^i=e^{-\lambda_i t}x^i$ (no summation), so that $\frac{\partial x^i}{\partial\tilde{x}^k}=e^{\lambda_i t}\delta_{ik}$ (no summation) and the transformed time-dependent 3-metric is $\tilde{g}_{ik}(\tilde{x},t)=e^{2\lambda_i t}\delta_{ik}$ (no summation). The corresponding 4-metric (in coordinates (t,x^i), *after removing the tilde*)

$$ds^2=-dt^2+e^{2\lambda_1 t}(dx^1)^2+e^{2\lambda_2 t}(dx^2)^2+e^{2\lambda_3 t}(dx^3)^2 \tag{1.60}$$

with $(\lambda_1,\lambda_2,\lambda_3)$ satisfying (1.59), is therefore a solution of the vacuum Gauss–Einstein equations (with $\Lambda=0$), that is, $R_{\mu\lambda}=0$. This can also easily be checked by direct computation. That this, the solution is non-trivial can be seen by computing, for example, the 1212-component of the Riemann curvature tensor. We get $R_{1212}=\lambda_1\lambda_2 e^{2(\lambda_1+\lambda_2)t}\neq 0$ (if $\lambda_1,\lambda_2\neq 0$).

The *complex* solution (1.53) is related to the so-called (*real*) Kasner ([8]) solutions in the following manner. Recall that the Kasner solutions for the vacuum Einstein equations $R_{\mu\lambda}=0$ are given by:

$$ds^2=t^{2\lambda_1}(dx^1)^2+t^{2\lambda_2}(dx^2)^2+t^{2\lambda_3}(dx^3)^2-t^{2\lambda_0}dt^2 \tag{1.61}$$

where the λ_α satisfy the two equations

$$\left.\begin{array}{r}\lambda_1+\lambda_2+\lambda_3=\lambda_0+1\\ \lambda_1^2+\lambda_2^2+\lambda_3^2=(\lambda_0+1)^2.\end{array}\right\} \tag{1.62}$$

If one takes $\lambda_0 = 0$, one gets the *standard* Kasner solution

$$ds^2 = -dt^2 + t^{2\lambda_1}(dx^1)^2 + t^{2\lambda_2}(dx^2)^2 + t^{2\lambda_3}(dx^3)^2 \quad (1.63)$$

with

$$\left.\begin{aligned} \lambda_1 + \lambda_2 + \lambda_3 = 1 \\ \lambda_1^2 + \lambda_2^2 + \lambda_3^2 = 1. \end{aligned}\right\} \quad (1.64)$$

A special solution of (1.64) is $\lambda_1 = \frac{2}{3}$, $\lambda_2 = \frac{2}{3}$, $\lambda_3 = -\frac{1}{3}$. The other obvious special solution $\lambda_1 = 1$, $\lambda_2 = \lambda_3 = 0$, i.e.,

$$ds^2 = -dt^2 + t^2(dx^1)^2 + (dx^2)^2 + (dx^3)^2 \quad (1.65)$$

is actually a flat metric as can be seen by the following transformation of coordinates: $\bar{t} = t \cosh x^1$, $\bar{x} = t \sinh x^1$, $\bar{x}^2 = x^2$, $\bar{x}^3 = x^3$, which transforms (1.65) into

$$ds^2 = -d\bar{t}^{\,2} + (d\bar{x}^1)^2 + (d\bar{x}^2)^2 + (d\bar{x}^3)^2.$$

If we replace t by $t = e^{\bar{t}}$ in (1.61), we get

$$ds^2 = e^{2\lambda_1\bar{t}}(dx^1)^2 + e^{2\lambda_2\bar{t}}(dx^2)^2 + e^{2\lambda_3\bar{t}}(dx^3)^2 - e^{2(\lambda_0+1)\bar{t}}(d\bar{t}\,)^2. \quad (1.66)$$

If we now take $\lambda_0 = -1$ in (1.62) and (1.66), we get precisely (1.60), with $\bar{t}$ instead of t,

$$ds^2 = e^{2\lambda_1\bar{t}}(dx^1)^2 + e^{2\lambda_2\bar{t}}(dx^2)^2 + e^{2\lambda_3\bar{t}}(dx^3)^2 - (d\bar{t}\,)^2$$

with

$$\left.\begin{aligned} \lambda_1 + \lambda_2 + \lambda_3 = 0 \\ \lambda_1^2 + \lambda_2^2 + \lambda_3^2 = 0. \end{aligned}\right\} \quad (1.59)$$

Consider now the *complex* solutions of (1.59). They are

$$\left.\begin{aligned} \lambda_1 = \frac{1}{2}(-1 + i\sqrt{3})\lambda_3 \\ \lambda_2 = \frac{1}{2}(-1 - i\sqrt{3})\lambda_3 \end{aligned}\right\} \quad (1.67)$$

where λ_3 can be *real* or *complex*. We shall choose $\lambda_3 = r$, a *real* number. Then $\lambda_1 = p + iq$, and its complex conjugate $\lambda_2 = p - iq$, with $p = -\frac{r}{2}$, $q = r\frac{\sqrt{3}}{2}$.

So that the non-zero (*complex*) metric tensor components for (1.60) are

$$\left.\begin{aligned} g_{11} &= e^{2(p+iq)t} \\ g_{22} &= e^{2(p-iq)t} \\ g_{33} &= e^{2rt} \\ g_{00} &= -1 \\ \text{with} \quad p &= -\frac{r}{2}, \quad q = r\frac{\sqrt{3}}{2}. \end{aligned}\right\} \tag{1.68}$$

We now make a *complex* coordinate transformation: $x^\alpha \mapsto \bar{x}^\alpha$ to obtain a *real* metric as follows:

$$\left.\begin{aligned} x^1 &= (a+ib)^{\frac{1}{2}}\bar{x}^1 + (c+id)^{\frac{1}{2}}\bar{x}^2 \\ x^2 &= (a-ib)^{\frac{1}{2}}\bar{x}^1 + (c-id)^{\frac{1}{2}}\bar{x}^2 \\ x^3 &= \bar{x}^3 \\ t &= \bar{t}. \end{aligned}\right\} \tag{1.69}$$

Here a, b, c, d are any *real* numbers, such that $\det(\frac{\partial x^\alpha}{\partial \bar{x}^\beta}) \neq 0$. The transformation matrix for (1.69) and its determinant are

$$\left(\frac{\partial x^\alpha}{\partial \bar{x}^\beta}\right) = \begin{pmatrix} (a+ib)^{\frac{1}{2}} & (c+id)^{\frac{1}{2}} & 0 & 0 \\ (a-ib)^{\frac{1}{2}} & (c-id)^{\frac{1}{2}} & 0 & 0 \\ 0 & 0 & 1 & 0 \\ 0 & 0 & 0 & 1 \end{pmatrix}, \tag{1.70}$$

$$\det\left(\frac{\partial x^\alpha}{\partial \bar{x}^\beta}\right) = [(a+ib)(c-id)]^{\frac{1}{2}} - [(a-ib)(c+id)]^{\frac{1}{2}}$$
$$= i2D$$

where

$$D = \operatorname{Im}[(a+ib)^{\frac{1}{2}}(c-id)^{\frac{1}{2}}]. \tag{1.71}$$

So the condition on a, b, c, d is that $D \neq 0$.

The transformed non-zero (real) metric tensor components are then

$$\bar{g}_{11} = \frac{\partial x^i}{\partial \bar{x}^1}\frac{\partial x^k}{\partial \bar{x}^1}\, g_{ik}$$
$$= \left(\frac{\partial x^1}{\partial \bar{x}^1}\right)^2 g_{11} + \left(\frac{\partial x^2}{\partial \bar{x}^1}\right)^2 g_{22}$$
$$= (a+ib)e^{2(p+iq)t} + (a-ib)e^{2(p-iq)t}$$
$$= 2e^{2pt}(a\cos 2qt - b\sin 2qt).$$

Similarly, $\bar{g}_{22} = 2e^{2pt}(c\cos 2qt - d\sin 2qt)$ and

$$\bar{g}_{12} = \frac{\partial x^i}{\partial \bar{x}^1}\,\frac{\partial x^k}{\partial \bar{x}^2}\, g_{ik} = \frac{\partial x^1}{\partial \bar{x}^1}\,\frac{\partial x^1}{\partial \bar{x}^2}\, g_{11} + \frac{\partial x^2}{\partial \bar{x}^1}\frac{\partial x^2}{\partial \bar{x}^2} g_{22}$$
$$= (a+ib)^{\frac{1}{2}}(c+id)^{\frac{1}{2}}e^{2(p+iq)t}$$
$$+ (a-ib)^{\frac{1}{2}}(c-id)^{\frac{1}{2}}e^{2(p-iq)t}$$
$$= 2e^{2pt}(A\cos 2qt - B\sin 2qt)$$

where

$$A = \operatorname{Re}[(a+ib)^{\frac{1}{2}}(c+id)^{\frac{1}{2}}] \tag{1.72}$$
$$B = \operatorname{Im}[(a+ib)^{\frac{1}{2}}(c+id)^{\frac{1}{2}}] \tag{1.73}$$

and

$$\bar{g}_{33} = g_{33} = e^{2rt}, \quad \bar{g}_{00} = g_{00} = -1.$$

We thus now have a *real* 4-metric $\bar{g}_{\alpha\beta}$ containing five *real* parameters a, b, c, d, r (in coordinates (t, x^i), *after removing the bar*):

$$\left.\begin{aligned} g_{11} &= 2e^{-rt}\big(a\cos(r\sqrt{3t}) - b\sin(r\sqrt{3t})\big) \\ g_{22} &= 2e^{-rt}\big(c\cos(r\sqrt{3t}) - d\sin(r\sqrt{3t})\big) \\ g_{12} &= 2e^{-rt}\big(A\cos(r\sqrt{3t}) - B\sin(r\sqrt{3t})\big) \\ g_{33} &= e^{2rt} \\ g_{00} &= -1. \end{aligned}\right\} \tag{1.74}$$

This is a solution of the vacuum Einstein equations $R_{\mu\lambda} = 0$, as can also be checked by direct computation.

The corresponding *real* vector field $\overline{X}$ is then given by $\overline{X} = \overline{X}^i(\bar{x})\frac{\partial}{\partial \bar{x}^i}$, where $\overline{X}^i(\bar{x}) = \frac{\partial \bar{x}^i}{\partial x^k} X^k(x)$. From (1.70), we have

$$\frac{\partial \bar{x}^i}{\partial x^k} = \begin{pmatrix} \dfrac{(c-id)^{1/2}}{i2D} & -\dfrac{(c+id)^{1/2}}{i2D} & 0 \\ -\dfrac{(a-ib)^{1/2}}{i2D} & \dfrac{(a+ib)^{1/2}}{i2D} & 0 \\ 0 & 0 & 1 \end{pmatrix},$$

so that,

$$\begin{aligned} \overline{X}^1(\bar{x}) &= \frac{(c-id)^{1/2}}{i2D}(-\lambda_1 x^1) - \frac{(c+id)^{1/2}}{i2D}(-\lambda_2 x^2) \\ &= -\frac{(p-iq)(c-id)^{1/2}}{i2D}\left[(a+ib)^{1/2}\bar{x}^1 + (c+id)^{1/2}\bar{x}^2\right] \\ &\quad + \frac{(p+iq)(c+id)^{1/2}}{i2D}\left[(a-ib)^{1/2}\bar{x}^1 + (c-id)^{1/2}\bar{x}^2\right] \\ &= \frac{\mathrm{Im}[(p-iq)(c+id)^{1/2}(a-ib)^{1/2}]}{\mathrm{Im}[(a+ib)^{1/2}(c-id)^{1/2}]}\,\bar{x}^1 - \frac{q(c^2+d^2)^{1/2}}{D}\bar{x}^2 \\ &= -q\bar{x}^1 - \frac{q(c^2+d^2)^{1/2}}{D}\bar{x}^2. \end{aligned} \tag{1.75}$$

Similarly, $\overline{X}^2(\bar{x}) = \frac{q(a^2+b^2)^{1/2}}{D}\bar{x}^1 + q\bar{x}^2$ and $\overline{X}^3(\bar{x}) = -r\bar{x}^3$. After *removing the bar* again, the *real* vector field X is given by

$$X = \left(-qx^1 - \frac{q(c^2+d^2)^{1/2}}{D}x^2\right)\frac{\partial}{\partial x^1} + \left(\frac{q(a^2+b^2)^{1/2}}{D}x^1 + qx^2\right)\frac{\partial}{\partial x^2} - rx^3\frac{\partial}{\partial x^3}. \tag{1.76}$$

We, thus, have a *real* solution (g, X) of (1.41)–(1.43) with $g = \delta$ and X given by (1.76).

A special case of (1.74) is given by $a = 1, b = c = 0, d = 1$ with $A = B = 1/\sqrt{2}, D = -1/\sqrt{2}$,

$$\left.\begin{aligned} \bar{g}_{11} &= 2e^{-r\tau}\cos(r\sqrt{3}\tau) \\ \bar{g}_{22} &= -2e^{-r\tau}\sin(r\sqrt{3}\tau) \\ \bar{g}_{12} &= \sqrt{2}e^{-r\tau}\left(\cos(r\sqrt{3}\tau) - \sin(r\sqrt{3}\tau)\right) \\ \bar{g}_{33} &= e^{2r\tau} \\ \bar{g}_{00} &= -1. \end{aligned}\right\} \tag{1.77}$$

This solution can also be checked by direct calculation.[d]

Notes on signature: The signature of the metric (1.77) is *non-Lorentzian* and the vector field X is real. A *Lorentzian* metric is obtained by changing the sign of $\bar{g}_{33}$, i.e., $\bar{g}_{33} = -e^{2r\tau}$. This can be achieved by taking $\tilde{x}^3 = i\bar{x}^3$ in (1.69). However, then, the corresponding vector field has to be *complex*. In both cases, the corresponding 3-metrics must have *indefinite* signatures.

[d]This was done on MAPLE, the symbolic computation language. See Appendix C.

1.9. A Solution of Mathematical Interest

We present now another solution of our equations which is mathematically interesting but turns out to be physically trivial. However, this solution illustrates the basic principle of our approach.

Again, suppose $\rho = \Lambda = 0$ and now we assume

$$h_{ik} = \lambda_i \lambda_k f \tag{1.78}$$

where λ_i are constants and f is a function of (x^i) to be determined. Then (1.51) is identically satisfied, whereas (1.50) implies

$$\frac{1}{2}(\lambda_i \lambda_1 f_{,1} + \lambda_i \lambda_2 f_{,2} + \lambda_i \lambda_3 f_{,3}) - \frac{1}{2}(\lambda_1^2 + \lambda_2^2 + \lambda_3^2) f_{,i} = 0. \tag{1.79}$$

We can satisfy (1.71) by assuming

$$\lambda_i f_{,k} = \lambda_k f_{,i} \tag{1.80}$$

which implies that

$$f(x) = F(u), \quad \text{where } u = \lambda_i x^i \tag{1.81}$$

and F is a function of the single variable u. Now, consider Eq. (1.52) in view of (1.78) and (1.80). It becomes

$$\frac{1}{2}(\mathcal{L}_X h)_{ik} = \frac{1}{4}\lambda_i \lambda_k f^2 \sum_j \lambda_j^2$$

or

$$\lambda_i \lambda_k f_{,\ell} X^\ell + \lambda_i \lambda_\ell f X^\ell_{,k} + \lambda_k \lambda_\ell f X^\ell_{,i} = \frac{1}{2}\lambda_i \lambda_k f^2 \sum_j \lambda_j^2. \tag{1.82}$$

Let us put

$$X^i = \frac{1}{2}\lambda_i G(u). \tag{1.83}$$

Then

$$h_{ik} = X^i_{,k} + X^k_{,i} = \lambda_i \lambda_k G',$$

so that

$$F = G' \tag{1.84}$$

and (1.82) becomes

$$\begin{aligned} (F'G + 2FG') \sum_j \lambda_j &= F^2 \sum_j \lambda_j^2, \\ \text{or} \quad [G''G + 2(G')^2] \sum_j \lambda_j &= (G')^2 \sum_j \lambda_j^2. \end{aligned} \tag{1.85}$$

We can satisfy (1.85) by taking, say,

$$\sum_j \lambda_j = \sum_j \lambda_j^2 = 1 \tag{1.86}$$

and by making sure that G satisfies the following equation

$$G''G + (G')^2 = 0, \tag{1.87}$$

the general solution of which is: $G(u) = \pm(au + b)^{\frac{1}{2}}$, where a, b are arbitrary constants.

A solution of (1.49)–(1.52) is thus provided by (g, X) where $g = \delta$ and $X = X^i(x)\delta/\delta x^i$ with $X^i(x) = \frac{1}{2}\lambda_i u^{1/2}$, where λ_i satisfy (1.86) and $u = \lambda_i x^i$. Next, we shall derive the corresponding 4-metric solution of the vacuum field equations (1.38)–(1.40) by transforming the 3-metric $g = \delta$ by the flow of X. The flow of X is given by the solution of the dynamical system:

$$\left.\begin{aligned} &\frac{dx^i}{dt} = \frac{1}{2}\lambda_i u^{\frac{1}{2}} = \lambda_i(\lambda_1 x^1 + \lambda_2 x^2 + \lambda_3 x^3)^{\frac{1}{2}} \\ &x^i(0) = x^i_0 \end{aligned}\right\} \tag{1.88}$$

together with (1.86). To solve (1.88), note that, according to (1.86) and (1.88) $\frac{du}{dt} = \lambda_i \frac{dx^i}{dt} = \frac{1}{2}(\Sigma_i \lambda_i^2) u^{\frac{1}{2}} = \frac{1}{2} u^{\frac{1}{2}}$ and $u(0) = u_0 = \lambda_i x_0^i$. This implies that $u^{\frac{1}{2}} = \frac{t}{4} + u_0^{\frac{1}{2}}$, and so

$$
\begin{aligned}
x^2(t) &= \int \frac{1}{2}\lambda_2 \left(\frac{t}{4} + u_0^{\frac{1}{2}}\right) dt = x_0^2 + \frac{1}{2}\lambda_2 u_0^{\frac{1}{2}} t + \frac{1}{16}\lambda_2 t^2, \\
x^3(t) &= \int \frac{1}{2}\lambda_3 \left(\frac{t}{4} + u_0^{\frac{1}{2}}\right) dt = x_0^3 + \frac{1}{2}\lambda_3 u_0^{\frac{1}{2}} t + \frac{1}{16}\lambda_3 t^2, \\
x^1(t) &= \frac{u}{\lambda_1} - \frac{\lambda_2}{\lambda_1} x^2(t) - \frac{\lambda_3}{\lambda_1} x^3(t) \\
&= \frac{1}{\lambda_1}\left(\frac{t}{4} + u_0^{\frac{1}{2}}\right)^2 - \frac{\lambda_2}{\lambda_1}\left(x_0^2 + \frac{1}{2}\lambda_2 u_0^{\frac{1}{2}} t + \frac{1}{16}\lambda_2 t^2\right) \\
&\quad - \frac{\lambda_3}{\lambda_1}\left(x_0^3 + \frac{1}{2}\lambda_3 u_0^{\frac{1}{2}} t + \frac{1}{16}\lambda_3 t^2\right) \\
&= x_0^1 + \frac{1}{2}\lambda_1 u_0^{\frac{1}{2}} + \frac{1}{16}\lambda_1 t^2.
\end{aligned}
$$

Under the flow of X, the point (x_0^i) is mapped into the point $(x^i(t))$. The flow generated by X is thus $\chi_t : x^i \mapsto \tilde{x}^i = x^i + \frac{1}{2}\lambda_i u^{\frac{1}{2}} t + \frac{1}{16}\lambda_i t^2$. The inverse transformation is: $\tilde{x}^i \mapsto x^i = \tilde{x}^i - \frac{1}{2}\lambda_i(\tilde{u}^{1/2} - \frac{t}{4})t - \frac{1}{16}\lambda_i t^2$, where $\tilde{u} = \lambda_i \tilde{x}^i$. In view of (1.86), the Jacobian matrix of the transformation turns out to be

$$
\left.
\begin{aligned}
&\frac{\partial x^i}{\partial \tilde{x}^k} = \delta_{ik} - \lambda_i \lambda_k \omega; \quad \omega = t/4(\lambda_i \tilde{x}^i)^{\frac{1}{2}} \\
\text{with} \quad &\det\left(\frac{\partial x^i}{\partial \tilde{x}^k}\right) = 1 - w.
\end{aligned}
\right\} \tag{1.89}
$$

Under the flow of X, the flat 3-metric $g_{ik}(x) = \delta_{ik}$ is transformed into $\tilde{g}_{ik}(\tilde{x}, t)$, according to:

$$
\tilde{g}_{ik}(\tilde{x}, t) = \frac{\partial x^\ell}{\partial \tilde{x}^i}\frac{\partial x^m}{\partial \tilde{x}^k}\delta_{\ell m}.
$$

Thus,

$$\tilde{g}_{ik}(\tilde{x}, t) = \delta_{ik} + \lambda_i \lambda_k \omega^2 - 2\lambda_i \lambda_k \omega. \tag{1.90}$$

And, we have a 4-metric solution $g_{\alpha\beta}$ of the vacuum field equations $R_{\alpha\beta} = 0$ (in coordinates (t, x^i), after *removing the tilde* in (1.90))

$$\begin{aligned} g_{\alpha\beta} &= \begin{pmatrix} \delta_{ik} + \lambda_i \lambda_k \omega^2 - 2\lambda_i \lambda_k \omega & 0 \\ 0 & -1 \end{pmatrix} \\ \omega &= t/4(\lambda_i x^i)^{\frac{1}{2}}; \quad \sum_i \lambda_i = \sum_i \lambda_i^2 = 1 \end{aligned} \tag{1.91}$$

with $\det(g_{\alpha\beta}) = -(1 - \omega)^2$. However, not only $R_{\alpha\beta} = 0$, but also $R_{\alpha\beta\gamma\delta} = 0$. The solution thus represents a flat space-time!

1.10. The Schwarzschild Solution

As the first example of a non-trivial flat 3-space solution of (1.41)–(1.43), consider a general spherically symmetric 3-metric g and a spherically symmetric vector field X in coordinates $(x^1 = \rho, x^2 = \theta, x^3 = \phi)$:

$$\begin{aligned} g &= g_{ik}dx^i dx^k = f(\rho)^2 d\rho^2 + \rho^2(d\theta^2 + \sin^2\theta d\phi^2), \\ X &= X^i \frac{\partial}{\partial x^i} = a(\rho)\frac{\partial}{\partial \rho}. \end{aligned} \tag{1.92}$$

Substituting (1.92) in (1.41)–(1.43), we obtain the following ordinary differential equations for the unknown functions $f(\rho)$ and $a(\rho)$

$$4\,\frac{\left(\frac{d}{d\rho}f(\rho)\right)a(\rho)}{f(\rho)\rho} = 0, \tag{1.93}$$

$$\begin{aligned} &4\,\frac{2\left(\frac{d}{d\rho}f(\rho)\right)\rho + (f(\rho))^3 - f(\rho) + 2\,\rho\left(\frac{d}{d\rho}f(\rho)\right)(a(\rho))^2(f(\rho))^2}{\rho^2(f(\rho))^3} \\ &+4\,\frac{2\,\rho(f(\rho))^3\left(\frac{d}{d\rho}a(\rho)\right)a(\rho) + (a(\rho))^2(f(\rho))^3}{\rho^2(f(\rho))^3} = 0, \end{aligned} \tag{1.94}$$

$$\begin{aligned} &-\Bigg(2\,\frac{d}{d\rho}f(\rho) + (f(\rho))^2\left(\frac{d^2}{d\rho^2}f(\rho)\right)(a(\rho))^2\rho \\ &\quad +3\left(\frac{d}{d\rho}f(\rho)\right)a(\rho)\,(f(\rho))^2\left(\frac{d}{d\rho}a(\rho)\right)\rho\Bigg) \times (f(\rho)\rho)^{-1} \\ &-\frac{a(\rho)(f(\rho))^3\left(\frac{d^2}{d\rho^2}a(\rho)\right)\rho + (f(\rho))^3\left(\frac{d}{d\rho}a(\rho)\right)^2\rho}{f(\rho)\rho} \end{aligned}$$

$$-\frac{2\left(\frac{d}{d\rho}f(\rho)\right)(a(\rho))^2(f(\rho))^2+2(f(\rho))^3\left(\frac{d}{d\rho}a(\rho)\right)a(\rho)}{f(\rho)\rho}=0, \tag{1.95}$$

$$-\frac{\left(\frac{d}{d\rho}f(\rho)\right)\rho+(f(\rho))^3-f(\rho)+(a(\rho))^2(f(\rho))^3}{(f(\rho))^3}$$

$$-\frac{2\,\rho(f(\rho))^3\left(\frac{d}{d\rho}a(\rho)\right)a(\rho)+\rho\left(\frac{d}{d\rho}f(\rho)\right)(a(\rho))^2(f(\rho))^2}{(f(\rho))^3}=0. \tag{1.96}$$

$$[\text{Eq. (1.96)}]\ (\sin(\theta))^2. \tag{1.97}$$

Equation (1.93) implies that, for non-trivial $a(\rho)$, $f(\rho)$ = const. (which we set equal to one). Equation (1.94) then becomes

$$2\left(\frac{d}{d\rho}a(\rho)\right)\frac{a(\rho)}{\rho}+\frac{a(\rho)^2}{\rho^2}=0 \tag{1.98}$$

whose solution is $a(\rho)=\frac{k}{\rho^{1/2}}$, k = const. The remaining Eqs. (1.95)–(1.97) are then identically satisfied.

Thus the flat 3-metric

$$\left.\begin{aligned} g = d\rho^2+\rho^2(d\theta^2+\sin^2\theta d\phi^2) \\ \text{and the vector field } X = k\rho^{-\frac{1}{2}}\frac{\partial}{\partial\rho} \end{aligned}\right\} \tag{1.99}$$

is the only solution of Eqs. (1.93)–(1.97).

1.10.1. *The Schwarzschild metric*

We now wish to derive the space-time 4-metric which corresponds to our solution (1.99), according to Theorem 1, by transforming the 3-metric g by the flow of the vector field X. In order to calculate the flow of X, it is convenient to transform X first to a simpler form by a simple change of coordinates. For example, the transformation $(\rho,\theta,\phi) \mapsto (R,\theta,\phi)$, where $\rho = (\frac{3}{2}kR)^{\frac{2}{3}}$ transforms the metric as well as the vector field in (1.99), into

$$\left.\begin{aligned} g &= k^2\left(\frac{3}{2}kR\right)^{-\frac{2}{3}} dR^2 + \left(\frac{3}{2}kR\right)^{\frac{4}{3}}(d\theta^2 + \sin^2\theta d\phi^2) \\ X &= \frac{\partial}{\partial R}. \end{aligned}\right\} \tag{1.100}$$

It is clear from the invariant form of Eqs. (1.41)–(1.43), that a change of coordinates provides another solution of the metric and the vector field. It can also be checked directly that (1.100) is indeed a solution of (1.41)–(1.43). And, of course, the 3-metric in (1.100) is still flat.

The flow φ_τ of X is given by: $\varphi_\tau : R \mapsto \widetilde{R} = R + \tau$, $\theta \mapsto \tilde{\theta} = \theta$, $\phi \mapsto \tilde{\phi} = \phi$. Therefore, according to Theorem 1, (1.25) corresponds to the *space-time* 4-*metric* (*in Gaussian normal coordinates* $(\tau, \widetilde{R}, \tilde{\theta}, \tilde{\phi})$):

$$\begin{aligned} ds^2 = -d\tau^2 &+ k^2\left[\frac{3}{2}k(\widetilde{R}-\tau)\right]^{-\frac{2}{3}} d\widetilde{R}^2 \\ &+ \left[\frac{3}{2}k(\widetilde{R}-\tau)\right]^{\frac{4}{3}}(d\tilde{\theta}^2 + \sin^2\tilde{\theta} d\tilde{\phi}^2). \end{aligned} \tag{1.101}$$

That (1.101) is a solution of the vacuum Einstein field equations can also be checked directly. In fact, (1.101) is the *Schwarzschild* solution in the so-called *Lemaitre coordinates* ([9–11]) if we take $k = (2m)^{\frac{1}{2}}$.

This can be seen by considering the following transformation:

$$(\tau, \widetilde{R}, \tilde{\theta}, \tilde{\phi}) \mapsto (t, r, \theta, \phi),$$

where

$$\left.\begin{aligned} \tau &= 2\left(\frac{r}{2m}\right)^{\frac{1}{2}} + 2m\log\left|\frac{\sqrt{r}-\sqrt{2m}}{\sqrt{r}+\sqrt{2m}}\right| - t = \tau(r,t) \\ \widetilde{R} &= \frac{2}{3}r^{\frac{3}{2}}(2m)^{-\frac{1}{2}} + \tau = \widetilde{R}(r,t) \\ \tilde{\theta} &= \theta, \quad \tilde{\phi} = \phi \end{aligned}\right\} \tag{1.102}$$

with the *inverse* transformation:

$$(t, r, \theta, \phi) \mapsto (\tau, \widetilde{R}, \tilde{\theta}, \tilde{\phi}),$$

where

$$\left.\begin{aligned} r &= (2m)^{\frac{1}{3}}\left[\frac{3}{2}(\widetilde{R}-\tau)\right]^{\frac{2}{3}} = r(\widetilde{R},\tau) \\ t &= 2\left(\frac{r}{2m}\right)^{\frac{1}{2}} + 2m\log\left|\frac{\sqrt{r}-\sqrt{2m}}{\sqrt{r}+\sqrt{2m}}\right| - \tau = t(\widetilde{R},\tau) \\ \theta &= \tilde{\theta}, \quad \phi = \tilde{\phi}. \end{aligned}\right\} \tag{1.103}$$

(1.102) or (1.103) transforms (1.101) into the Schwarzschild metric:

$$ds^2 = -\left(1-\frac{2m}{r}\right)dt^2 + \left(1-\frac{2m}{r}\right)^{-1}dr^2 + r^2(d\theta^2 + \sin^2\theta d\phi^2). \tag{1.104}$$

Incidentally, we have thus proved a version of Birkoff's Theorem in our formalism, namely, that *the only spherically symmetric solution of* (1.41)–(1.43) *leads to the Schwarzschild space-time.*

One might think that if (X, g) is a solution of (1.41)–(1.43), then adding a Killing vector field of g to X would again give us another solution. In fact, one has the following proposition.

Proposition 1. *Let (X, g) be a solution and X_0 a Killing vector field of g. Then, $(X + X_0, g)$ is also a solution provided $[X, X_0] = 0$.*

Proof. Let $\tilde{X} = X + X_0$. Then, $\mathcal{L}_{\tilde{X}} g = \mathcal{L}_X g$ and $\mathcal{L}_{\tilde{X}}(\mathcal{L}_{\tilde{X}} g) = \mathcal{L}_X(\mathcal{L}_X g) + \mathcal{L}_{X_0}(\mathcal{L}_X g)$. Now, $[\mathcal{L}_{X_0}, \mathcal{L}_X] = \mathcal{L}_{[X_0, X]}$. Hence, $[X_0, X] = 0$ implies that $\mathcal{L}_{X_0}(\mathcal{L}_X g) = \mathcal{L}_X(\mathcal{L}_{X_0} g) = 0$, and therefore, we have also $\mathcal{L}_{\tilde{X}}(\mathcal{L}_{\tilde{X}} g) = \mathcal{L}_X(\mathcal{L}_X g)$. □

However, this does not generate a new space-time 4-metric in view of the following Proposition.

Proposition 2. (X, g) *and* $(X + X_0, g)$, *where* $[X, X_0] = 0$, *generate equivalent space-time* 4-*metrics.*

Proof. Since X commutes with X_0, the flow of $X + X_0$ is a composition (as maps) of the flow of X and the flow of X_0. Since X_0 is Killing, the flow of X_0 has no effect on the metric g. Therefore, the effect of the flow of $X + X_0$ on g is the same as that of X. □

For example, consider the solution (1.99) together with the two Killing vectors X_0 and X_1 of the 3-metric:

$$
\begin{aligned}
g\colon \quad ds^2 &= d\rho^2 + \rho^2(d\theta^2 + \sin^2\theta d\phi^2), \\
X &= (2m/\rho)^{\frac{1}{2}} \frac{\partial}{\partial \rho}, \\
X_0 &= \sin\phi \frac{\partial}{\partial \theta} + \cot\theta \cos\phi \frac{\partial}{\partial \phi}, \\
X_1 &= \sin\theta \cos\phi \frac{\partial}{\partial \rho} + (\cos\theta \cos\phi/\rho) \frac{\partial}{\partial \theta} \\
&\quad - (\operatorname{cosec}\theta \sin\phi/\rho) \frac{\partial}{\partial \phi}.
\end{aligned}
$$

Here, the Killing vector field X_0 corresponds to rotational isometry, whereas X_1 corresponds to translational isometry. $[X, X_0] = 0$, but $[X, X_1] \neq 0$. $(X + X_0, g)$ is again a solution, but $(X + X_1, g)$ is not. The 4-metrics corresponding to (X, g) and $(X + X_0, g)$ are equivalent.

Our definition of a *physical observer* was that of a vector field X on (M^3, g) such that $0 < g(X, X) < 1$. Note that the vector field $X = [\frac{2m}{\rho}]^{\frac{1}{2}} \frac{\partial}{\partial \rho}$ in (1.99), therefore, ceases to be *physical* when $\rho \leq 2m$ or when $R \leq \frac{4}{3}m$, even though X has a *mathematical* singularity only at $\rho = 0$. $\rho = 2m$ is, of course, the Schwarzschild horizon corresponding to $r = 2m$ in Schwarzschild coordinates (t, r, θ, ϕ).

1.11. From Space-Time 4-Metric to 3-Metric and 3-Vector Field

There exists a *converse* to *Theorem* 1.

Theorem 2. *Suppose a space-time* 4*-metric solution of the vacuum Einstein field equations has the form*

$$ds^2 = -d\tau^2 + \tilde{g}_{ik}(\tau, \tilde{x}^i)d\tilde{x}^i d\tilde{x}^k$$

in the Gaussian normal coordinates $(\tau, \tilde{x}^i)$, *where* $\tilde{g}_{ik}(\tau, \tilde{x}^i)$ *is the transformed* τ*-dependent metric by the flow* φ_τ *of some* 3*-vector field* X *acting on a* 3*-metric* $g_{ik}(x^i)$. *Then* (g_{ik}, X) *is a solution of our Eqs.* (1.41)–(1.43).

Proof. By assumption, $\tilde{g} = \tilde{\varphi}_\tau(g)$ satisfies Eqs. (1.46)–(1.48). Again, from (1.45), one has

$$\tilde{\varphi}_\tau(\mathcal{L}_X K)|_{\tau=0} = \frac{\partial}{\partial\tau}(\tilde{\varphi}_\tau(K))\bigg|_{\tau=0}.$$

Since $\tilde{\varphi}_0 =$ Identity, (1.46)–(1.48) implies (1.41)–(1.43). □

Corollary. *In particular, if a space-time* 4*-metric solution of the vacuum Einstein equations has the form:*

$$ds^2 = -d\tau^2 + \tilde{g}_{ik}(\tilde{x}^1 - \tau, \tilde{x}^2, \tilde{x}^3)d\tilde{x}^i d\tilde{x}^k \tag{1.105}$$

in the Gaussian normal coordinates $(\tau, \tilde{x}^i)$, *then* $(g_{ik} = \tilde{g}_{ik}(x^1, x^2, x^3), X = 1\frac{\partial}{\partial x^1})$ *is a solution of* (1.41)–(1.43).

This is because the flow of $X = 1\frac{\partial}{\partial x^1}$ is simply $\varphi_\tau : x^1 \mapsto \tilde{x}^1 = x^1 + \tau, x^2 \mapsto \tilde{x}^2 = x^2, x^3 \mapsto \tilde{x}^3 = x^3$.

As an example of *Theorem* 2, we shall now illustrate how to obtain the equivalent 3-metric g and 3-vector field X from a space-time 4-metric which satisfies the vacuum Einstein field equations and which has the specific form given by (1.105). The first example will be the Schwarzschild black hole metric and the procedure below will be helpful when we consider next the Kerr black hole in our 3-dimensional formalism.

Consider the Schwarzschild metric gx in coordinates $x = [x1, x2, x3, x4]$

$$gx = \begin{pmatrix} \dfrac{x1^2}{x1^2 - 2mx1} & 0 & 0 & 0 \\ 0 & x1^2 & 0 & 0 \\ 0 & 0 & x1^2 \sin^2 x2 & 0 \\ 0 & 0 & 0 & -1 + \dfrac{2m}{x1} \end{pmatrix} \tag{1.106}$$

and, in anticipation of (1.103), make a coordinate transformation of the form

$$\begin{gathered} [x1, x2, x3, x4] \mapsto [y1, y2, y3, y4] \\ x1 = F(y1 - y4), \quad x2 = y2, \\ x3 = y3, \quad x4 = H(F(y1 - y4)) - y4, \end{gathered} \tag{1.107}$$

where the functions F and H would be determined by imposing suitable conditions in order to bring the metric into the appropriate form. Then, the transformed 4-metric components $gy\alpha\beta$ $(\alpha, \beta = 1, 2, 3, 4)$ are (see Appendix D)

$$\begin{aligned} gy11 = {} & -(\mathrm{D}(F)(y1 - y4))^2[-(F(y1 - y4))^2 \\ & + (\mathrm{D}(H)(F(y1 - y4)))^2(F(y1 - y4))^2 \\ & - 4(\mathrm{D}(H)(F(y1 - y4)))^2 mF(y1 - y4) \\ & + 4(\mathrm{D}(H)(F(y1 - y4)))^2 m^2]/F(y1 - y4)(F(y1 - y4) - 2m), \end{aligned} \tag{1.108}$$

$$gy12 = 0, \tag{1.109}$$

$$gy13 = 0, \tag{1.110}$$

$$gy22 = F(y1 - y4)^2, \tag{1.111}$$

$$gy23 = 0, \tag{1.112}$$

$$gy33 = (F(y1 - y4))^2(\sin(y2))^2, \tag{1.113}$$

$$\begin{aligned}
gy14 = {} & \mathrm{D}(F)(y1 - y4)[-\mathrm{D}(F)(y1 - y4)(F(y1 - y4))^2 \\
& + (\mathrm{D}(H)(F(y1 - y4)))^2\mathrm{D}(F)(y1 - y4)(F(y1 - y4))^2 \\
& - 4\,(\mathrm{D}(H)(F(y1 - y4)))^2\mathrm{D}(F)(y1 - y4)mF(y1 - y4) \\
& + 4\,(\mathrm{D}(H)(F(y1 - y4)))^2\mathrm{D}(F)(y1 - y4)m^2 \\
& + \mathrm{D}(H)(F(y1 - y4))(F(y1 - y4))^2 \\
& - 4\mathrm{D}(H)(F(y1 - y4))mF(y1 - y4) \\
& + 4\mathrm{D}(H)(F(y1 - y4))m^2]/F(y1 - y4)(F(y1 - y4) - 2m),
\end{aligned} \tag{1.114}$$

$$gy24 = 0, \tag{1.115}$$

$$gy34 = 0, \tag{1.116}$$

$$\begin{aligned}
gy44 = {} & -[-(\mathrm{D}(F)(y1 - y4))^2(F(y1 - y4))^2 \\
& + (\mathrm{D}(H)(F(y1 - y4)))^2(\mathrm{D}(F)(y1 - y4))^2(F(y1 - y4))^2 \\
& - 4(\mathrm{D}(H)(F(y1 - y4)))^2(\mathrm{D}(F)(y1 - y4))^2mF(y1 - y4) \\
& + 4(\mathrm{D}(H)(F(y1 - y4)))^2(\mathrm{D}(F)(y1 - y4))^2m^2 \\
& + 2\mathrm{D}(H)(F(y1 - y4))\mathrm{D}(F)(y1 - y4)(F(y1 - y4))^2 \\
& - 8\mathrm{D}(H)(F(y1 - y4))\mathrm{D}(F)(y1 - y4)mF(y1 - y4) \\
& + 8\mathrm{D}(H)(F(y1 - y4))\mathrm{D}(F)(y1 - y4)m^2 + (F(y1 - y4))^2 \\
& - 4mF(y1 - y4) + 4m^2]/F(y1 - y4)(F(y1 - y4) - 2m),
\end{aligned} \tag{1.117}$$

where D is the derivative operator.

To reduce it to the appropriate Gaussian normal form we need to impose the conditions

$$gy14 = 0, \tag{1.118}$$

$$gy44 = -1. \tag{1.119}$$

One can solve (1.118)–(1.119) for the derivatives $\mathrm{D}(F)$, $\mathrm{D}(H)$ in terms of $F(y1-y4)$ to obtain

$$\mathrm{D}(F)(y1-y4) = \sqrt{2}\sqrt{\frac{m}{F(y1-y4)}}, \tag{1.120}$$

$$\mathrm{D}(H)(F(y1-y4)) = \sqrt{2}\sqrt{\frac{m}{F(y1-y4)}}\left(1-2\,\frac{m}{F(y1-y4)}\right)^{-1}. \tag{1.121}$$

Substituting (1.120)–(1.121) back into (1.108)–(1.117) we obtain the desired Gaussian normal form for the Schwarzschild metric (provided F and H satisfy (1.120)–(1.121))

$$\left.\begin{aligned} &gy14 = gy24 = gy34 = 0, \quad gy44 = -1 \\ &gy11 = 2\,\frac{m}{F(y1-y4)} \\ &gy12 = gy13 = gy23 = 0 \\ &gy22 = (F(y1-y4))^2 \\ &gy33 = (F(y1-y4))^2(\sin(y2))^2. \end{aligned}\right\} \tag{1.122}$$

According to the *Corollary* of *Theorem* 2 the above 4-metric is generated by the 3-metric gz in coordinates $z = [z1, z2, z3]$ (one simply replaces $y1 - y4$ by $z1$, $y2$ by $z2$ and $y3$ by $z3$)

$$gz = \begin{pmatrix} 2m/F(z1) & 0 & 0 \\ 0 & F(z1)^2 & 0 \\ 0 & 0 & F(z1)^2(\sin(z2))^2 \end{pmatrix} \tag{1.123}$$

and the 3-vector field

$$Xz = 1\,\frac{\partial}{\partial z1}. \tag{1.124}$$

One can now easily solve (1.120) to find that $F(z1) = ((3/2)(2m)^{1/2}z1)^{2/3}$ (apart from a constant), and thus recover essentially (1.100).

However, we do not need to solve (1.120) *explicitly* for $F(z1)$ in order to obtain (1.99). We simply make a sort of *inverse* transformation of the coordinates $[z1, z2, z3]$ back to the *original* coordinates. In other words, a transformation

$$\left.\begin{aligned} [z1, z2, z3] &\mapsto [w1, w2, w3] \\ z1 = F^{-1}(w1), \quad z2 &= w2, \quad z3 = w3 \end{aligned}\right\} \tag{1.125}$$

where $z1 = F^{-1}(w1)$ is the inverse function of $w1 = F(z1)$, so that $F(F^{-1}(w1)) = w1$ and $\frac{dF^{-1}}{dw1} = 1/(\frac{dF}{dz1}) = 1/(2m/w1)^{1/2}$ from (1.120). Under (1.125), (1.123)–(1.124) is, therefore, transformed into

$$\left.\begin{aligned} &gw11 = 2m\left(\frac{dF^{-1}}{dw1}\right)^2 / F(F^{-1}(w1)) = 1 \\ &gw12 = gw13 = gw23 = 0 \\ &gw22 = F(F^{-1}(w1))^2 = w1^2 \\ &gw33 = F(F^{-1}(w1))^2(\sin(w2))^2 = w1^2(\sin(w2))^2, \end{aligned}\right\} \tag{1.126}$$

$$Xw = \left(1 \Big/ \left(\frac{dF^{-1}}{dw1}\right)\right)\frac{\partial}{\partial w1} = (2m/w1)^{1/2}\frac{\partial}{\partial w1} \tag{1.127}$$

which is essentially (1.99).

1.12. The Kerr Solution

As for the second example of Theorem 2, we shall now demonstrate that the axially symmetric stationary Kerr solution [12] of the vacuum Einstein equations can also be brought into the appropriate Gaussian normal form by a procedure similar to the one outlined in the previous section, and thus can also be formulated in terms of a 3-metric and a 3-vector field on a 3-manifold. Consider the Kerr metric in Boyer–Lindquist coordinates $[x1, x2, x3, x4]$

$$\begin{aligned}
gx11 &= \frac{x1^2 + a^2(\cos(x2))^2}{x1^2 - 2mx1 + a^2},\\
gx12 &= gx13 = gx23 = gx14 = gx24 = 0,\\
gx22 &= x1^2 + a^2(\cos(x2))^2,\\
gx33 &= \frac{((x1^2 + a^2)^2 - (x1^2 - 2mx1 + a^2)a^2(\sin(x2))^2)(\sin(x2))^2}{x1^2 + a^2(\cos(x2))^2},\\
gx34 &= -2\,\frac{amx1\,(\sin(x2))^2}{x1^2 + a^2(\cos(x2))^2},\\
gx44 &= -1 + 2\,\frac{mx1}{x1^2 + a^2(\cos(x2))^2},
\end{aligned} \tag{1.128}$$

and make a coordinate transformation of the form:

$$\left.\begin{aligned}
&[x1, x2, x3, x4] \mapsto [y1, y2, y3, y4],\\
&x1 = F(y1 - y4, y2), \quad x2 = y2, \quad x3 = K(y1 - y4, y2) + y3,\\
&x4 = H(F(y1 - y4, y2)) - y4.
\end{aligned}\right\} \tag{1.129}$$

We choose the unknown function F, K, H in such a way that the following conditions are satisfied for the transformed metric components. (The expressions for the transformed metric components are too long to reproduce here. See Appendix D.)

$$gy14 = gy24 = gy34 = 0, \quad gy44 = -1. \tag{1.130}$$

These equations can now be solved for the following derivatives of F, K, H, in terms of $F(y1 - y4)$ to give

$$\begin{aligned}
&\mathrm{D}[1](F)(y1 - y4, y2) \\
&\quad = \frac{\sqrt{2}\sqrt{m}\sqrt{F(y1 - y4, y2)a^2 + (F(y1 - y4, y2))^3}}{(F(y1 - y4, y2))^2 + a^2(\cos(y2))^2}, \qquad (1.131)\\
&D[1](K)(y1 - y4, y2) \\
&\quad = 2amF(y1 - y4, y2)/[((F(y1 - y4, y2))^2 + a^2(\cos(y2))^2) \\
&\quad\quad \times ((F(y1 - y4, y2))^2 - 2mF(y1 - y4, y2) + a^2)], \qquad (1.132)\\
&\mathrm{D}(H)(F(y1 - y4, y2)) \\
&\quad = \frac{\sqrt{2}\sqrt{m}\sqrt{F(y1 - y4, y2)a^2 + (F(y1 - y4, y2))^3}}{(F(y1 - y4, y2))^2 - 2mF(y1 - y4, y2) + a^2}. \qquad (1.133)
\end{aligned}$$

Here $\mathrm{D}[1](F)$, $\mathrm{D}[1](K)$ are the first partial derivatives of F, K with respect to the first argument $y1 - y4$, respectively. Note that there are no conditions on the first partial derivatives of F, K with respect to the second argument $y2$.

The above conditions (1.131)–(1.133) are not only necessary but also sufficient for (1.130) to hold. In principle, therefore, one can solve (1.131) for $F(y1 - y4, y2)$ and, then, (1.132)–(1.133) to determine $K(y1 - y4, y2)$, $H(F(y1 - y4, y2))$. Unfortunately, (1.131) cannot be integrated *explicitly* as it involves elliptic integrals.

Let us assume the above conditions (1.131)–(1.133). The components of gy then have the appropriate Gaussian normal form given by (1.105). They depend on functions of $(y1-y4, y2)$ and $y2$ only. Therefore, according to the *Corollary* of *Theorem* 2, gy is generated by the

3-metric gz in coordinates $[z1, z2, z3]$ (where one replaces $y1 - y4$ by $z1$, $y2$ by $z2$ and $y3$ by $z3$ in $gyik$ $(i, k = 1, 2, 3)$ to get $gzik$) and the 3-vector field

$$Xz = 1\frac{\partial}{\partial z1} \tag{1.134}$$

$gzik$ $(i, k = 1, 2, 3)$ contain only $F(z1, z2)$, $K(z1, z2)$ and their first partial derivatives.

To obtain an *explicit* form of the 3-metric we make an "inverse" transformation of the coordinates $[z1, z2, z3]$ (analogous to the Schwarzschild case) back to the "original" coordinates as follows:

$$\begin{aligned} &[z1, z2, z3] \mapsto [w1, w2, w3] \\ &z1 = F^{-1}(w1, w2), \quad z2 = w2, \\ &z3 = w3 - K(F^{-1}(w1, w2), w2) \end{aligned} \tag{1.135}$$

where F^{-1} is the inverse function of $w1 = F(z1, z2)$ *with respect to the first variable*, that is, $F(F^{-1}(w1, w2), w2) = w1$.

Under (1.135) the vector field Xz is transformed into

$$\begin{aligned} Xw = {} & \left[1 \Big/ \frac{\partial F^{-1}}{\partial w1}(w1, w2)\right] \frac{\partial}{\partial w1} \\ & + [\mathrm{D}[1](K)(F^{-1}(w1, w2), w2)] \frac{\partial}{\partial w3}. \end{aligned} \tag{1.136}$$

The transformed components $gwik$ contain $F(F^{-1}(w1, w2), w2)$ and only the following derivatives:

$$\begin{aligned} &\frac{\partial F^{-1}}{\partial w1}(w1, w2), \quad \frac{\partial F^{-1}}{\partial w2}(w1, w2), \\ &\mathrm{D}[1](K)(F^{-1}(w1, w2), w2), \\ &\mathrm{D}[1](F)(F^{-1}(w1, w2), w2), \\ &\mathrm{D}[2](F)(F^{-1}(w1, w2), w2). \end{aligned}$$

These can be expressed *explicitly* in terms of $w1, w2$ as follows. From (1.131) and (1.132), we have

$$F(F^{-1}(w1, w2), w2) = w1, \tag{1.137}$$

$$\begin{aligned} &\mathrm{D}[1](K)(F^{-1}(w1, w2), w2) \\ &\quad = \frac{\partial K}{\partial z1} = 2\,\frac{amw1}{(w1^2 + a^2(\cos(w2))^2)(w1^2 + a^2 - 2mw1)}, \end{aligned} \tag{1.138}$$

$$\begin{aligned} &\frac{\partial F^{-1}}{\partial w1}(w1, w2) \\ &\quad = 1 \Big/ \frac{\partial F}{\partial z1} = 1/2\,\frac{(w1^2 + a^2(\cos(w2))^2)\sqrt{2}}{\sqrt{m}\sqrt{w1(w1^2 + a^2)}}. \end{aligned} \tag{1.139}$$

Now, from (1.131), by integrating once

$$\frac{\partial w1}{\partial z1} = \frac{\sqrt{2}\sqrt{m}\sqrt{w1(w1^2 + a^2)}}{w1^2 + a^2(\cos(w2))^2},$$

we have

$$z1 = \frac{1}{(2m)^{1/2}} \int \frac{w1^2 + a^2(\cos(w2))^2}{\sqrt{w1(w1^2 + a^2)}} dw1,$$

apart from a function of $w2$, which we set equal to zero. Therefore,

$$\frac{\partial z1}{\partial w2} = -\frac{2a^2 \sin(w2)\cos(w2)}{(2m)^{1/2}} \int \frac{dw1}{\sqrt{w1(w1^2 + a^2)}}$$

or

$$\frac{\partial F^{-1}}{\partial w2}(w1, w2) = -\frac{2a^2 \sin(w2)\cos(w2)}{(2m)^{1/2}} EL(w1), \tag{1.140}$$

where $EL(w1)$ is a standard elliptic integral. And, again from (1.131),

$$\begin{aligned} &\mathrm{D}[1](F)(F^{-1}(w1, w2), w2) \\ &\quad = \frac{\partial F}{\partial z1} = \frac{\sqrt{2}\sqrt{m}\sqrt{w1(w1^2 + a^2)}}{w1^2 + a^2(\cos(w2))^2}. \end{aligned} \tag{1.141}$$

To calculate $\mathrm{D}[2](F)(F^{-1}(w1, w2), w2) = \frac{\partial F}{\partial z2}$, we differentiate both sides of (1.137) with respect to $w2$, to get

$$\begin{aligned}
0 = \frac{\partial w1}{\partial w2} &= \frac{\partial F}{\partial z1}\frac{\partial F^{-1}}{\partial w2}(w1, w2) + \frac{\partial F}{\partial z2}\frac{\partial z2}{\partial w2} \\
&= \left[\frac{\sqrt{2}\sqrt{m}\sqrt{w1(w1^2 + a^2)}}{w1^2 + a^2(\cos(w2))^2}\right] \\
&\quad \times \left[-\frac{a^2 \sin(w2)\cos(w2)EL(w1)\sqrt{2}}{\sqrt{m}}\right] + \frac{\partial F}{\partial z2} \times 1
\end{aligned}$$

so that

$$\begin{aligned}
&\mathrm{D}[2](F)(F^{-1}(w1, w2), w2) \\
&\quad = 2\frac{a^2 \sin(w2)\cos(w2)EL(w1)\sqrt{w1(w1^2 + a^2)}}{w1^2 + a^2(\cos(w2))^2}.
\end{aligned} \tag{1.142}$$

Finally, the *explicit expressions for the* 3*-metric* gw and the 3*-vector field* Xw, which generate the Kerr solution, are:

$$\begin{aligned}
gw11 = &\frac{w1^2 + a^2(\cos(w2))^2}{w1^2 + a^2 - 2mw1} \\
&- 2\frac{mw1(w1^2 + a^2)(w1^2 + a^2(\cos(w2))^2 - 2mw1)}{(w1^2 + a^2(\cos(w2))^2)(w1^2 + a^2 - 2mw1)^2}
\end{aligned} \tag{1.143}$$

$$gw12 = gw23 = 0, \tag{1.144}$$

$$gw13 = -2\frac{a(\sin(w2))^2\sqrt{2}m^{3/2}w1\sqrt{w1^3 + a^2w1}}{(w1^2 + a^2(\cos(w2))^2)(w1^2 + a^2 - 2mw1)}, \tag{1.145}$$

$$gw22 = w1^2 + a^2(\cos(w2))^2, \tag{1.146}$$

$$gw33 = \left(w1^2 + a^2 - 2mw1 + 2\frac{mw1(w1^2 + a^2)}{w1^2 + a^2(\cos(w2))^2}\right)(\sin(w2))^2, \tag{1.147}$$

$$Xw = \left[\frac{\sqrt{2}\sqrt{m}\sqrt{w1^3 + a^2 w1}}{w1^2 + a^2(\cos(w2))^2}\right]\frac{\partial}{\partial w1} + \left[2\,\frac{amw1}{(w1^2 + a^2(\cos(w2))^2)(w1^2 + a^2 - 2mw1)}\right]\frac{\partial}{\partial w3}. \tag{1.148}$$

One can now verify (Appendix E) that the above (g, X) is indeed a solution of our Eqs. (1.41)–(1.43).

Note that, (1.143)–(1.145) reduce to the Schwarzschild case (1.126)–(1.127) when $a \to 0$. When $m \to 0$, we obtain

$$gw = \begin{bmatrix} \frac{w1^2 + a^2(\cos(w2))^2}{w1^2 + a^2} & 0 & 0 \\ 0 & w1^2 + a^2 \times (\cos(w2))^2 & 0 \\ 0 & 0 & (w1^2 + a^2) \times (\sin(w2))^2 \end{bmatrix}, \tag{1.149}$$

$$Xw = 0. \tag{1.150}$$

The above 3-metric (1.149) is flat, and therefore (1.149)–(1.150) is equivalent to the flat space-time.

If $a \neq 0$, the 3-metric (1.143)–(1.145) is *not* flat, in contrast to (1.126), the 3-metric in the Schwarzschild case. Its scalar curvature is given by

$$Rw = \frac{2a^2 mw1(3(\cos(w2))^2 - 1)}{w1^2 + a^2(\cos(w2))^2} \tag{1.151}$$

and the 3-vector field Xw in (1.148) has the length

$$g(X, X) = gw(Xw, Xw) = \frac{2mw1}{w1^2 + a^2(\cos(w2))^2}. \tag{1.152}$$

Thus X ceases to be "*physical*" when $g(X, X) = 1$, i.e.,

$$\frac{2mw1}{w1^2 + a^2(\cos(w2))^2} = 1 \tag{1.153}$$

which corresponds to the so-called *stationary limit* of the Kerr black hole.

1.13. The Maxwell Equations

Our basic approach to evolution in a 3-dimensional manifold can also be applied to the Maxwell equations.

We first write the Maxwell equations on a space-time

$$\left.\begin{aligned} \operatorname{curl}\mathbf{E} &= -\frac{\partial \mathbf{B}}{\partial t} \\ \operatorname{curl}\mathbf{H} &= 4\pi\mathbf{j} + \frac{\partial \mathbf{D}}{\partial t} \\ \operatorname{div}\mathbf{D} &= 4\pi\rho \\ \operatorname{div}\mathbf{B} &= 0 \end{aligned}\right\} \tag{1.154}$$

for the electromagnetic variables $\mathbf{E}, \mathbf{H}, \mathbf{B}, \mathbf{D}, \mathbf{j}, \rho$ (the usual electric, magnetic, fields and inductions, current and charges) in terms of 3-*dimensional time-dependent differential forms* on a 3-manifold (M^3, g) as follows. Define the following (in some local coordinates (x^i) of (M^3, g)):

1-forms, $\alpha, \beta \in \Lambda^1(M^3)$:

$$\alpha = \sum_i \alpha_i dx^i, \quad \mathbf{E} = (\alpha_1, \alpha_2, \alpha_3),$$

$$\beta = \sum_i \beta_i dx^i, \quad \mathbf{H} = (\beta_1, \beta_2, \beta_3).$$

2-forms, $\gamma, \delta, \epsilon \in \Lambda^2(M^3)$:

$$\gamma = \sum_{i<j} \gamma_{ij} dx^i \wedge dx^j, \quad \mathbf{B} = (\gamma_{23}, \gamma_{31}, \gamma_{12}),$$

$$\delta = \sum_{i<j} \delta_{ij} dx^i \wedge dx^j, \quad \mathbf{D} = (\delta_{23}, \delta_{31}, \delta_{12}),$$

$$\epsilon = \sum_{i<j} \epsilon_{ij} dx^i \wedge dx^j, \quad \mathbf{j} = (\epsilon_{23}, \epsilon_{31}, \epsilon_{12}).$$

3-form: $\eta \in \Lambda^3(M^3)$:

$$\eta = \rho \, dx^1 \wedge dx^2 \wedge dx^3.$$

Note that these 3-dimensional forms are time-dependent with the time t as a parameter. Using the exterior derivative operator d the Maxwell equations (1.154) can then be written as:

$$\left.\begin{aligned} d\alpha &= -\frac{\partial \gamma}{\partial t} \\ d\beta &= 4\pi\epsilon + \frac{\partial \delta}{\partial t} \\ d\delta &= 4\pi\eta \\ d\gamma &= 0. \end{aligned}\right\} \tag{1.155}$$

As we did in the case of Gauss–Einstein equations, let us now *replace the time-derivative* $\frac{\partial}{\partial t}$ *in* (1.155) *by the Lie-derivative* $\mathcal{L}_X$ *with respect* to some vector field X on M^3. Then (1.155) becomes

$$d\alpha = -\mathcal{L}_X \gamma, \tag{1.156}$$

$$d\beta = 4\pi\epsilon + \mathcal{L}_X \delta, \tag{1.157}$$

$$d\delta = 4\pi\eta, \tag{1.158}$$

$$d\gamma = 0. \tag{1.159}$$

Here, $\alpha, \beta, \gamma, \delta, \epsilon, \eta$ are to be regarded now as (*time-independent*) purely 3-dimensional differential forms on M^3. Equations (1.156)–(1.159) are thus analogous to (1.38)–(1.40), corresponding to the Gauss–Einstein equations, and are to be considered as a set of equations for the forms $\alpha, \beta, \gamma, \delta, \epsilon, \eta$ and the vector field X. Every solution of (1.156)–(1.159) determines uniquely a solution of (1.155) and, thus, of (1.154).[e] One simply has to transform the time-independent

[e] Again, the converse need not be true.

differential form α, for example, by the flow χ_t of X to obtain a time-dependent form: $\tilde{\alpha} = \tilde{\chi}_t(\alpha)$, etc. The vector field X thus provides the temporal evolution of the electromagnetic fields.

The above equations take a surprisingly simple (and elegant) form if one uses the relationship between the Lie derivative $\mathcal{L}_X$ and the exterior derivative d and the contraction operator i_X relative to X:

$$\mathcal{L}_X = i_X \cdot d + d \cdot i_X \tag{1.160}$$

to obtain from (1.156) and (1.159)

$$d\alpha = -(i_X \cdot d + d \cdot i_X)\gamma = -d(i_X\gamma)$$

or

$$d\Gamma = 0 \quad \text{where } \Gamma = \alpha + i_X\gamma. \tag{1.161}$$

Similarly, from (1.87) and (1.88)

$$d\beta = 4\pi\epsilon + (i_X \cdot d + d \cdot i_X)\delta = 4\pi(\epsilon + i_X\eta) + d(i_X\delta)$$

or

$$d\Omega = 4\pi\Theta \quad \text{where } \Omega = \beta - i_X\delta, \quad \Theta = \epsilon + i_X\eta. \tag{1.162}$$

Thus, Eqs. (1.156)–(1.159) become

$$\left.\begin{aligned} d\Gamma &= 0 \\ d\Omega &= 4\pi\Theta \end{aligned}\right\} \tag{1.163}$$

where

$$\left.\begin{aligned} \Gamma &= \alpha + i_X\gamma \\ \Omega &= \beta - i_X\delta \\ \Theta &= \epsilon + i_X\eta. \end{aligned}\right\} \tag{1.164}$$

In vacuum, $\epsilon = \eta = \Theta = 0$; thus, the corresponding vacuum equations are

$$\left.\begin{aligned} d\Gamma &= 0 \\ d\Omega &= 0. \end{aligned}\right\} \tag{1.165}$$

The continuity equation follows by taking the exterior derivative of the second equation in (1.163), i.e., $d\Theta = 0$. From (1.164)

$$0 = d(\epsilon + i_X \eta) = d\epsilon + (\mathcal{L}_X - i_X \cdot d)\eta.$$

Since η is a 3-form on a 3-manifold, $d\eta = 0$. Thus

$$\mathcal{L}_X \eta + d\epsilon = 0 \tag{1.166}$$

which is the continuity equation relative to the vector field (observer) X if we keep in mind that the Lie-derivative $\mathcal{L}_X$ corresponds to the time-derivative $\frac{\partial}{\partial t}$ and $d\epsilon$ is the divergence of the current $\mathbf{j}$.

References

[1] G. D. Mostow, *Strong Rigidity of Locally Symmetric Spaces*, Ann. Math. Series (Princeton Univ. Press, Princeton, 1973).

[2] S. Helgason, *Differential Geometry and Symmetric Spaces* (Academic, New York, 1972).

[3] S. Kobayashi and K. Nomizu, *Foundations of Differential Geometry*, Vol. 1 (Interscience, New York, 1983).

[4] D. V. Widder, *The Heat Equation* (Academic Press, New York, 1975).

[5] A. Lichnerowicz, *Théories relativistes de la gravitation et de l'électromagnetisme* (Masson, Paris, 1955).

[6] J. S. Synge, *Relativity, the General Theory* (North-Holland, Amsterdam, 1960).

[7] A. E. Fisher and J. E. Marsden, *J. Math. Phys.* **13** (1972) 546–568.

[8] E. Kasner, *American J. Math.* **43** (1921) 217.

[9] G. Lemaitre, L'univers en expansion, *Ann. Soc. Sci. Bruxelles I A* **53** (1933) 51.

[10] A. Z. Petrov, *Einstein Spaces* (Pergamon Press, Oxford, 1969).

[11] D. K. Sen, *Class. Quant. Gravity* **12** (1995) 553–577.

[12] D. K. Sen, *J. Math. Phys.* **41** (2000) 7556–7572.

2. MATTER

2.0. Introduction

The basic constituents of all matter, namely, the elementary particles, have the following salient features.

The massive ones, that is, those with non-zero rest mass, can be classified into two basic categories: the lighter *leptons* and the heavier *baryons*.. In each category, there is a *stable* particle, the lepton *electron* and the baryon *proton*. The *unstable* leptons and baryons all spontaneously decay eventually into the stable ones.

Then, there are the so-called zero rest mass particles, the *photon* and the *neutrino*, which are essentially carriers of energy and vehicles of interaction between the particles.

The so-called standard model has had considerable success as a unified theory of all elementary particles. Together with the Higgs boson and Higgs mechanism, we are able to explain the existence and masses of several new bosons.

Nevertheless, it cannot be considered as a complete and fully satisfactory theory of all elementary particles. It cannot, for example, explain intrinsically why the proton is about 1836 times heavier than the electron.

In the standard model, the phenomenon of neutrino oscillation ([1]) requires that neutrinos have non-zero mass.

In 1957, Heisenberg ([2, 3]) tried to formulate (without much success) a unified theory of all elementary particles starting from a nonlinear 4-component spinor equation with a built-in fundamental constant.

Here, we ([4, 5]) suggest that massless 2-component Weyl neutrinos, instead of 4-component spinors, are probably more fundamental than previously thought. We consider a composite system consisting of a massless positively oriented 2-component Weyl neutrino and a massless negatively oriented 2-component Weyl neutrino with a certain specific symmetry-breaking interaction between the two.

We assume that the observable physical particles manifest as energy states of the resulting 4-component system. A simple quantum mechanical treatment shows that such a model should exhibit 2-fold branching and energy defects, which could then be interpreted as formation of particles of non-zero rest mass.

Such a model can also provide a qualitative, alternative non-standard explanation of the different flavors of a massless 4-component neutrino and thus, of neutrino oscillation without assuming a neutrino mass.

2.1. The Photon and the Weyl Neutrinos

A long time ago, de Broglie ([6]) and Jordan ([7]) have tried to construct a neutrino theory of light based on the former's idea of "fusion". Both these authors tried to explain the photon as a combination of two 4-component neutrinos of essentially the same kind. Pryce ([8]) and Barbour *et al.* ([9]) however showed that such a photon would be *longitudinally* polarized. For a historical review of the "neutrino theory of light", see [10].

We first show that the photon is intimately connected not to the 4-component Dirac neutrino, but to the 2-component positively and negatively oriented Weyl neutrinos.

The Maxwell equations in vacuum can be written as ([11, 12])

$$df = 0, \quad d * f = 0 \tag{2.1}$$

where f is the electromagnetic differential 2-form, d is the exterior derivative operator and $*f$ is the Hodge dual of f. We shall use space-time coordinates $x_\mu (= x, y, z, ict)$, $\mu = 1, 2, 3, 4$, so that the metric tensor in a Minkowski space-time is $\delta_{\mu\nu}$ and consider only proper Lorentz transformations. Then, in components, the Hodge dual of f is given by

$$*f_{\mu\nu} = \frac{1}{2}\varepsilon_{\mu\nu\alpha\beta} f_{\alpha\beta} \tag{2.2}$$

where $\varepsilon_{\mu\nu\alpha\beta}$ is the completely anti-symmetric Levi–Civita symbol.

In general, for a p-form w on an n-dimensional pseudo-Riemannian manifold, one has

$$* * w = (-1)^{p(n-p)} w \tag{2.3}$$

so that $**f = f$ for the electromagnetic 2-form f on the 4-dimensional Minkowski space-time. It is thus possible to decompose f in an invariant manner into a self-dual part f^s and an anti-self-dual part f^a, as follows

$$f = f^s + f^a, \quad f^s = \frac{1}{2}(f + *f), \quad f^a = \frac{1}{2}(f - *f) \tag{2.4}$$

where $*f^s = f^s$, $*f^a = -f^a$. The Maxwell equations (2.1) are then equivalent to

$$f = f^s + f^a, \quad df^s = 0, \quad df^a = 0. \tag{2.5}$$

For an anti-symmetric tensor, the self-duality condition (in components) $f^s_{\mu\nu} = *f^s_{\mu\nu}$ implies that

$$f^s_{12} = f^s_{34}, \quad f^s_{23} = f^s_{14}, \quad f^s_{31} = f^s_{24} \tag{2.6}$$

so that it has only three independent components. Now, a self-dual (or an anti-self-dual) anti-symmetric tensor cannot by itself describe a physical electromagnetic field, because in that case there should exists a 4-vector $A_\mu = (\mathbf{A}, i\phi)$, such that $f^s_{\nu\mu} = \partial_\mu A_\nu - \partial_\nu A_\mu$. No such 4-vector with three real components and one imaginary component can exist if $f^s_{\mu\nu}$ were to be self-dual, because (2.6) would then imply that

$$\partial_y A_x - \partial_x A_y = (1/ic)\partial_t A_z - i\partial_z \phi. \tag{2.7}$$

Consider, however, a 2-component quantity $\varphi = \binom{\varphi_1}{\varphi_2}$ and a 4-vector

$$\psi^s = \begin{pmatrix} \varphi_1 \\ i\varphi_1 \\ -\varphi_2 \\ i\varphi_2 \end{pmatrix} \tag{2.8}$$

and set $f^s_{\mu\lambda} = \partial_\lambda \psi^s_\mu - \partial_\mu \psi^s_\lambda$. Then, the self-duality condition (2.6) gives basically two equations

$$f^s_{12} = \partial_2\varphi_1 - i\partial_1\varphi_1 = -\partial_4\varphi_2 - i\partial_3\varphi_2 = f^s_{34}, \tag{2.9}$$

$$f^s_{23} = i\partial_3\varphi_1 + \partial_2\varphi_2 = \partial_4\varphi_1 - i\partial_1\varphi_2 = f^s_{14}. \tag{2.10}$$

The third condition in (2.6) gives again (2.10). Multiplying (2.9) by i and (2.10) by $-i$, we get two equations for (φ_1, φ_2):

$$\left.\begin{aligned} \partial_1\varphi_1 + i\partial_2\varphi_1 - \partial_3\varphi_2 + i\partial_4\varphi_2 = 0 \\ \partial_1\varphi_2 - i\partial_2\varphi_2 + \partial_3\varphi_1 + i\partial_4\varphi_1 = 0 \end{aligned}\right\} \tag{2.11}$$

which by introducing the Pauli spin matrices

$$\sigma_1 = \begin{pmatrix} 0 & 1 \\ 1 & 0 \end{pmatrix}, \quad \sigma_2 = \begin{pmatrix} 0 & -i \\ i & 0 \end{pmatrix}, \quad \sigma_3 = \begin{pmatrix} 1 & 0 \\ 0 & -1 \end{pmatrix} \tag{2.12}$$

can be written as a single equation $(k = 1, 2, 3)$

$$(\sigma_k\partial_k + i\partial_4)\varphi = 0 \tag{2.13}$$

which is nothing but the Weyl equation for the 2-component left-handed neutrino ν_L. It is easily seen that if ψ^s is to be a 4-vector, φ must transform as a 2-component spinor ([4]).

Following a similar procedure for the anti-self-dual case, let

$$\psi^a = \begin{pmatrix} -\chi_2 \\ i\chi_1 \\ \chi_1 \\ i\chi_2 \end{pmatrix} \tag{2.14}$$

where $\chi = \binom{\chi_1}{\chi_2}$ is 2-component quantity and set $f^a_{\mu\lambda} = \partial_\lambda \psi^a_\mu - \partial_\mu \psi^a_\lambda$.

Then, the anti-self-duality condition for $f^a_{\mu\lambda}$ implies that

$$f^a_{24} = f^a_{13}, \quad f^a_{34} = f^a_{21}, \quad f^a_{14} = f^a_{32} \tag{2.15}$$

which in turn implies two equations for (χ_1, χ_2).

$$\left.\begin{aligned} \partial_1\chi_1 - i\partial_2\chi_2 + \partial_3\chi_2 + i\partial_4\chi_1 = 0 \\ -\partial_1\chi_2 + i\partial_2\chi_1 + \partial_3\chi_1 + i\partial_4\chi_2 = 0. \end{aligned}\right\} \tag{2.16}$$

Introducing the matrices

$$\varrho_1 = \begin{pmatrix} 1 & 0 \\ 0 & -1 \end{pmatrix}, \quad \varrho_2 = \begin{pmatrix} 0 & -i \\ i & 0 \end{pmatrix}, \quad \varrho_3 = \begin{pmatrix} 0 & 1 \\ 1 & 0 \end{pmatrix}, \tag{2.17}$$

(2.16) can be written as

$$(\varrho_k\partial_k + i\partial_4)\chi = 0. \tag{2.18}$$

Equation (2.18) differs from Eq. (2.13) only in the respect that the set ϱ_k differs from the Pauli matrices σ_k only in the interchange of the indices 1 and 3. As a result, while the Pauli matrices satisfy

$$\sigma_1\sigma_2 = i\sigma_3, \quad \sigma_2\sigma_3 = i\sigma_1, \quad \sigma_3\sigma_1 = i\sigma_2, \tag{2.19}$$

the set ϱ_k satisfy

$$\varrho_1\varrho_2 = -i\varrho_3, \quad \varrho_2\varrho_3 = -i\varrho_1, \quad \varrho_3\varrho_1 = -i\varrho_2. \tag{2.20}$$

Therefore, Eq. (2.18) describes a 2-component right-handed neutrino ν_R.

Conversely, if φ and χ satisfy (2.13) and (2.18), respectively, then $f_{\mu\lambda} = f^s_{\mu\lambda} + f^a_{\mu\lambda}$ with $f^s_{\mu\lambda} = \partial_\lambda\psi^s_\mu - \partial_\mu\psi^s_\lambda$ and $f^a_{\mu\lambda} = \partial_\lambda\psi^a_\mu - \partial_\mu\psi^a_\lambda$, where ψ^s and ψ^a are given by (2.8) and (2.14), would automatically satisfy the Maxwell equations in vacuum ([8]). In other words, the photon γ can be regarded in some sense as a "fusion" of a left-handed and a right-handed neutrino ν_L and ν_R.

This suggests that the left-handed and right-handed neutrinos are perhaps the fundamental constituents of all elementary particles and that one should consider composite ν_L–ν_R models with some fundamental interaction between the two. Ideally, such a model should be treated in the framework of quantum field theory. However, as a first step, in this work, we shall consider a simple, heuristic quantum mechanical model. This will give us an idea of what we can expect in a rigorous quantum field theoretical model.

2.2. A Neutrino Theory of Matter

2.2.1. *Composite ν_L–ν_R system without interaction*

Consider first a composite ν_L–ν_R system without interaction. The Hamiltonian of a left-handed (massless) neutrino ν_L described by (2.13) is given by $H_L = -ic\hbar(\sigma \cdot \nabla)$ with $\sigma = (\sigma_1, \sigma_2, \sigma_3)$ and its eigenfunctions φ_{E_L} satisfy

$$\left.\begin{aligned} H_L \varphi_{E_L} &= E_L \varphi_{E_L} \\ \text{or} \quad (\sigma \cdot \nabla)\varphi_{E_L} &= (iE_L/c\hbar)\varphi_{E_L}. \end{aligned}\right\} \tag{2.21}$$

From now on, we shall adopt the conventional units in which $c = \hbar = 1$.

The solutions of (2.21) are well-known:

$$\varphi_{E_L}(\mathbf{x}, \mathbf{p}) = a(\mathbf{p})e^{i\mathbf{x}\cdot\mathbf{p}}$$

where

$$\mathbf{p}^2 \equiv p_1^2 + p_2^2 + p_3^2 = E_L^2$$

and

$$a(\mathbf{p}) = \begin{pmatrix} p_1 - ip_2 \\ E_L - p_3 \end{pmatrix}. \tag{2.22}$$

They describe a left-handed neutrino ν_L with energy E_L.

The spectrum of H_L is not discrete and hence, the eigenfunctions have to be normalized by the delta-function.

$$\langle \varphi_{E_L}(\mathbf{x}, \mathbf{p}) | \varphi_{E'_L}(\mathbf{x}, \mathbf{p}') \rangle = a(\mathbf{p})^\dagger a(\mathbf{p}')\delta(\mathbf{p} - \mathbf{p}'). \tag{2.23}$$

This may give rise to questions of rigor and technical difficulties in what follows. One can always enclose the neutrino in a cubical box of length L, in which case the eigenfunctions would be

$$\varphi_{E_L}(\mathbf{x}, \mathbf{n}) = a(\mathbf{n})e^{i(2\pi/L)\mathbf{x}\cdot\mathbf{n}},$$

where $\mathbf{n} = (n_1, n_2, n_3)$, $n_1^2 + n_2^2 + n_3^2 = E_L^2$, with integer n_i, and therefore, discrete energy values. We shall not follow this procedure and proceed as if we are dealing with a discrete problem.

The eigenfunctions are also ∞-fold degenerate, since $\mathbf{p}$ can take any value on the energy shell. We shall remove this degeneracy by integrating $\varphi_{E_L}(\mathbf{x}, \mathbf{p})$ over the energy shell $S^2_{E_L} : p_1^2 + p_2^2 + p_3^2 = E_L^2$ which is a 2-sphere of radius E_L. We get (with a slight abuse of notation)

$$\varphi_{E_L}(\mathbf{x}) = \int_{S^2_{E_L}} a(\mathbf{p})e^{i\mathbf{x}\cdot\mathbf{p}}dS_p \tag{2.24}$$

as a surface integral over $S^2_{E_L}$ (see later). Furthermore, we shall suppose that $\varphi_{E_L}(\mathbf{x})$ are normalized.

Similarly, the corresponding eigenfunctions for the right-handed neutrino ν_R with Hamiltonian $H_R = -i(\rho \cdot \nabla)$ with energy E_R are given by

$$\left.\begin{aligned} \chi_{E_R}(\mathbf{y}, \mathbf{q}) &= b(\mathbf{q})e^{i\mathbf{y}\cdot\mathbf{q}} \\ \text{where} \quad q_1^2 + q_2^2 + q_3^2 &= E_R^2 \\ \text{and} \quad b(\mathbf{q}) &= \begin{pmatrix} q_3 - iq_2 \\ E_R - q_1 \end{pmatrix}. \end{aligned}\right\} \tag{2.25}$$

We use $\mathbf{y}$ for the coordinate of ν_R and note that $b(\mathbf{q})$ differs from $a(\mathbf{p})$ by an interchange of the indices 1 and 3. A similar integration over the energy shell $S^2_{E_R} : q_1^2 + q_2^2 + q_3^2 = E_R^2$ gives

$$\chi_{E_R}(\mathbf{y}) = \int_{S^2_{E_R}} b(\mathbf{q})e^{i\mathbf{y}\cdot\mathbf{q}}dS_q. \tag{2.26}$$

Let $\mathcal{H}_\mathcal{L}$ and $\mathcal{H}_\mathcal{R}$ be the respective Hilbert spaces for ν_L and ν_R. Consider now a composite ν_L–ν_R system without interaction given by the tensor product $\mathcal{H} = \mathcal{H}_\mathcal{L} \otimes \mathcal{H}_\mathcal{R}$ with the Hamiltonian $H_o = H_L + H_R$ (or more precisely, $H_L \otimes I + I \otimes H_R$) ([13]). Since $H_o = -i(\sigma \cdot \nabla_L) - i(\rho \cdot \nabla_R)$, where ∇_L, ∇_R act on $\mathbf{x}, \mathbf{y}$, respectively, and thus does not depend on $\mathbf{x}$ and $\mathbf{y}$ explicitly, not only

$$\Psi_1(\mathbf{x}, \mathbf{y}) = \varphi_{E_L}(\mathbf{x}) \otimes \chi_{E_R}(\mathbf{y}) \tag{2.27}$$

but also

$$\Psi_2(\mathbf{x}, \mathbf{y}) = \varphi_{E_R}(\mathbf{x}) \otimes \chi_{E_L}(\mathbf{y}) \tag{2.28}$$

are both eigenfunctions of H_o with energy $E_L + E_R$. Thus, the composite system ν_L–ν_R behaves like a system of identical particles even though ν_l is not identical to ν_R. We remark that if the universe were spatially non-orientable, ν_L would be indistinguishable from ν_R. But in an orientable universe, they would be distinct particles. We thus have a 2-fold degeneracy similar to the He-atom.

2.2.2. *Composite ν_L–ν_R system with interaction*

Consider now an interaction $V(\mathbf{x}, \mathbf{y})$ between ν_L and ν_R of the form

$$V(\mathbf{x}, \mathbf{y}) = F(|\mathbf{x}|)H(|\mathbf{x}| - |\mathbf{y}|) + F(|\mathbf{y}|)H(|\mathbf{y}| - |\mathbf{x}|) \tag{2.29}$$

where F is a function to be specified and H is the Heaviside function. Thus

$$V(\mathbf{x},\mathbf{y}) = \left.\begin{array}{ll} F(|\mathbf{x}|) & \text{if } |\mathbf{x}| > |\mathbf{y}| \\ F(|\mathbf{y}|) & \text{if } |\mathbf{x}| < |\mathbf{y}|. \end{array}\right\} \tag{2.30}$$

Note that $V(\mathbf{x},\mathbf{y}) = V(\mathbf{y},\mathbf{x})$. The combined Hamiltonian $H = H_o + V(\mathbf{x},\mathbf{y})$ with total energy, say, E_{Tot} satisfies an equation of the form:

$$[H_L + H_R + V(\mathbf{x},\mathbf{y})]\Psi = E_{\text{Tot}}\Psi. \tag{2.31}$$

We can now follow an approximation (i.e., perturbation) procedure similar to that of the He-atom ([14]). (Although it would be perhaps more appropriate to use the Lippmann–Schwinger equation for the continuous case.) Thus, as a first approximation the total energy is given by

$$E_{\text{Tot}} = E_L + E_R + (J \pm K) \tag{2.32}$$

where

$$\begin{aligned} J &= \langle \Psi_1 | V(\mathbf{x},\mathbf{y}) | \Psi_1 \rangle \\ &= \langle \varphi_{E_L}(\mathbf{x}) \otimes \chi_{E_R}(\mathbf{y}) | V(\mathbf{x},\mathbf{y}) | \varphi_{E_L}(\mathbf{x}) \otimes \chi_{E_R}(\mathbf{y}) \rangle, \quad (2.33) \\ K &= \langle \Psi_2 | V(\mathbf{x},\mathbf{y}) | \Psi_1 \rangle \\ &= \langle \varphi_{E_R}(\mathbf{x}) \otimes \chi_{E_L}(\mathbf{y}) | V(\mathbf{x},\mathbf{y}) | \varphi_{E_L}(\mathbf{x}) \otimes \chi_{E_R}(\mathbf{y}) \rangle. \quad (2.34) \end{aligned}$$

We shall now suppose that V is such that J is non-positive, i.e., $J(E_L, E_R) = -\Gamma(E_L, E_R)$, where $\Gamma(E_L, E_R) \geq 0$ for $E_L, E_R \geq 0$ (see example later), so that

$$E_{\text{Tot}} = E_L + E_R - (\Gamma \pm |K|). \tag{2.35}$$

The two branches of E_{Tot} would then have energy defects and the zeros of these two branches where E_{Tot} become positive again could then be identified with the Baryon and Lepton masses E_{Bar} and E_{Lep}.

2.2.3. *A reduced 1-dimensional model*

In order to evaluate J and K for any specific model, we need first to evaluate the surface integrals (2.24) and (2.26) over the energy shells $S^2_{E_L}$ and $S^2_{E_R}$.

Setting $p_3 = \pm\sqrt{E_L^2 - (p_1^2 + p_2^2)}$ and introducing the polar coordinates (r, θ) in the p_1–p_2 plane, we get

$$\varphi_{E_L}(\mathbf{x}) = \begin{pmatrix} \varphi^1_{E_L}(\mathbf{x}) \\ \varphi^2_{E_L}(\mathbf{x}) \end{pmatrix},$$

where

$$\varphi^1_{E_L}(\mathbf{x}) = \int_0^{E_L} \left(\int_0^{2\pi} e^{i(ra(\theta)-\theta)} d\theta \right) 2 \frac{\cos(x_3\sqrt{{E_L}^2 - r^2}) E_L r^2}{\sqrt{{E_L}^2 - r^2}} dr \tag{2.36a}$$

$$\varphi^2_{E_L}(\mathbf{x}) = \int_0^{E_L} \left(\int_0^{2\pi} e^{ira(\theta)} d\theta \right) \times 2 \frac{\left(E_L \cos(x_3\sqrt{{E_L}^2 - r^2}) - i\sqrt{{E_L}^2 - r^2} \sin(x_3\sqrt{{E_L}^2 - r^2}) \right) E_L r}{\sqrt{{E_L}^2 - r^2}} dr. \tag{2.36b}$$

Here, $a(\theta) = x_1 \cos(\theta) + x_2 \sin(\theta)$.

Unfortunately, the above integrals cannot be evaluated in closed form.

Consider, therefore, a reduced 1-dimensional model with $x_1 = x_2 = 0$. Then $a(\theta) = 0$, so that (setting $x_3 = x$)

$$\varphi_{E_L}(x) = \begin{pmatrix} 0 \\ 4\,\dfrac{(E_L \sin(E_L x)x - i\sin(E_L x) + ixE_L\cos(E_L x))\,E_L\pi}{x^2} \end{pmatrix} \tag{2.37}$$

$\varphi_{E_L}(x)$ is square integrable, that is

$$\langle \varphi_{E_L}(x) | \varphi_{E_L}(x) \rangle = \int_{-\infty}^{\infty} \varphi_{E_L}(x) \dagger \varphi_{E_L}(x) dx = \left(\frac{64}{3}\right) \pi^3 E_L^5. \tag{2.38}$$

Figure 2 shows how $\varphi_{E_L}(x) \dagger \varphi_{E_L}(x)$ behaves with x.

The normalizing factor for $\varphi_{E_L}(x)$ is thus $N_{E_L} = \left(\left(\frac{64}{3}\right)\pi^3 E_L^5\right)^{-1/2}$.

From now on, we shall suppose that $\varphi_{E_L}(x)$ is normalized.

Similarly, setting $y_1 = y_2 = 0$, $y_3 = y$ we get from (2.26)

$$\chi_{E_R}(y) = \begin{pmatrix} \dfrac{4i\,(\sin(E_R y) - E_R y\cos(E_R y))\,\pi\, E_R}{y^2} \\ 4\,\dfrac{\sin(E_R y)\pi\, {E_R}^2}{y} \end{pmatrix} \tag{2.39}$$

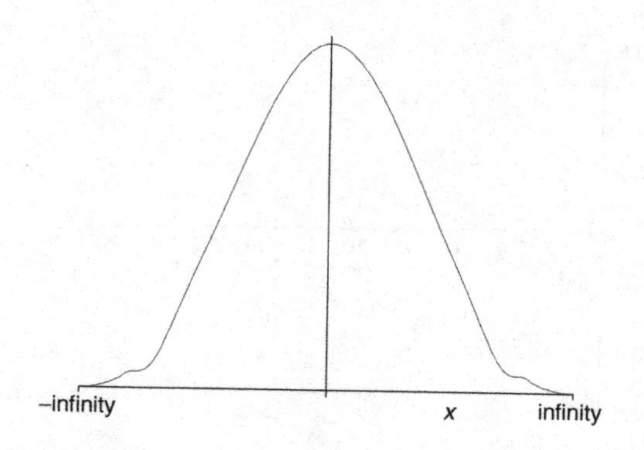

Fig. 2. $\varphi_{E_L}(x) \dagger \varphi_{E_L}(x)$ as a function of x.

and

$$\langle \chi_{E_R}(y) | \chi_{E_R}(y) \rangle = \int_{-\infty}^{\infty} \chi_{E_R}(y) \dagger \chi_{E_R}(y) dy = \left(\frac{64}{3}\right) \pi^3 E_R^5 \quad (2.40)$$

so that the normalizing factor for $\chi_{E_R}(y)$ is also $N_{E_R} = \left((\frac{64}{3})\pi^3 E_R^5\right)^{-1/2}$. Again, suppose that from now on, $\chi_{E_R}(y)$ is normalized.

According to (2.33)

$$J = \iint \varphi_{E_L}(x) \dagger \varphi_{E_L}(x) V(x,y) \chi_{E_R}(y) \dagger \chi_{E_R}(y) dx dy. \quad (2.41)$$

Using a mean-value theorem for integrals, we can write

$$J = \varphi_{E_L}(\xi) \dagger \varphi_{E_L}(\xi) \chi_{E_R}(\eta) \dagger \chi_{E_R}(\eta) \iint V(x,y) dx dy \quad (2.42)$$

where $-\infty < \xi, \eta < \infty$.

Now *assume* that V is given by (2.30), where the function F is, for $x > 0$ (see Fig. 3):

$$F(x) = e^{-\mu x} \ln(x). \quad (2.43)$$

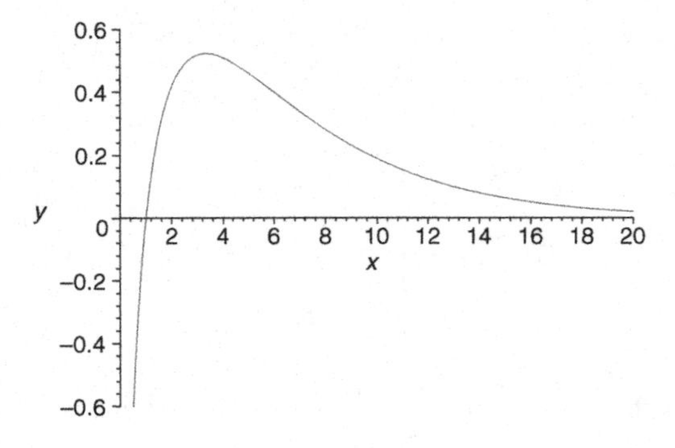

Fig. 3. $F(x)$ for $x > 0$.

Here, $\mu > 0$ is some fundamental interaction constant. Then,

$$\left.\begin{aligned}\iint V(x,y)dxdy &= 4\int_0^\infty dy\left[\int_0^y F(y)dx + \int_y^\infty F(x)dx\right]\\ &= -8(\ln(\mu)+\gamma-1)/\mu^2 \equiv -\Gamma_o.\end{aligned}\right\} \quad (2.44)$$

Here, $\gamma = 0.5772156649\ldots.$ (Euler constant). Hence, $\Gamma_o > 0$ if $\mu > e^{1-\gamma} = 1.526205112\ldots.$ We shall suppose that this condition is satisfied, so that $J = -\Gamma \leq 0$, where

$$\Gamma(E_L, E_R) = \varphi_{E_L}(\xi) \dagger \varphi_{E_L}(\xi)\chi_{E_R}(\eta) \dagger \chi_{E_R}(\eta)\Gamma_o. \quad (2.45)$$

Similarly,

$$K(E_L, E_R) = -\varphi_{E_R}(\vartheta) \dagger \varphi_{E_L}(\vartheta)\chi_{E_L}(\zeta) \dagger \chi_{E_R}(\zeta)\Gamma_o \quad (2.46)$$

where $-\infty < \vartheta, \zeta < \infty$.

In order to obtain a qualitative idea of how the two branches of E_{Tot} behave as a function of E_L and E_R, consider the case where $E_L = E_R = E$. Then, $\Gamma = |K|$ and for the branch $E_{\mathrm{Tot}} = 2E - (\Gamma + |K|)$ we have, from (2.37), (2.39) and (2.45),

$$\left.\begin{aligned}E_{\mathrm{Tot}} = \frac{2E-3}{4}\,&\frac{E^2\xi^2+1-(\cos(E\xi))^2-2\sin(E\xi)\xi E\cos(E\xi)}{\pi E^3\xi^4}\\ &\times\frac{3}{4}\,\frac{E^2\eta^2+1-(\cos(E\eta))^2-2\sin(E\eta)\eta E\cos(E\eta)}{\pi E^3\eta^4}\Gamma_o.\end{aligned}\right\} \quad (2.47)$$

Figure 4 is a plot of E_{Tot} as a function of E for some specific values of ξ, η, Γ_o. It shows that, in general, for values $0 < E < E-$, E_{Tot} is positive; it becomes zero at $E = E-$; it is then negative when $E- < E < E+$. It is positive again when $E > E+$.

How can one interpret this somewhat strange behavior of E_{Tot} as a function of E?

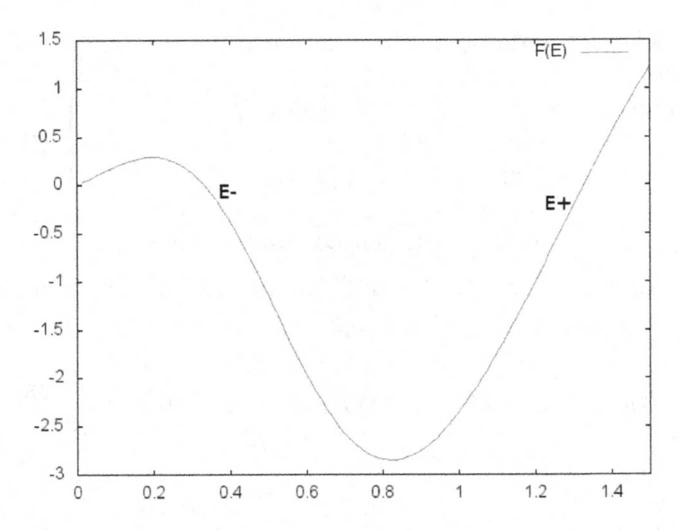

Fig. 4. E_{Tot} as a function of E.

A possible interpretation is as follows.

Now, imagine that we start with a free ν_L and a free ν_R — both of energy E. The free system has a mirror symmetry. If we now switch on the above-mentioned interaction, the symmetry is broken because of the nature of the Heaviside function in the interaction. Then, depending on if $0 < E < E-$, the composite system is a 4-component Dirac neutrino of energy $E_{\text{Tot}} > 0$. If, however $E- < E < E+$, the composite system would contribute to vacuum energy $E_{\text{Tot}} < 0$. If $E > E+$, we have a positive energy defect of $2E+$. This can be interpreted as the formation of a particle of rest mass equal to $2E+$ and E_{Tot} now representing the pure kinetic energy of the formed particle.

The rest mass $2E+$ would depend on the value of the fundamental constant μ.

A similar analysis of the other branch (with $E_L \neq E_R$) would lead to another particle with rest mass less than the above case.

2.2.4. *Conclusion*

We have thus shown that a simple quantum mechanical analysis of a composite ν_L–ν_R model with a specific symmetry-breaking interaction suggests a possible formation of particles of non-zero rest mass from 2-component Weyl neutrinos of sufficiently high energy. In this interactive model, in a model universe filled with Weyl neutrinos ν_L and ν_R with energy spectrum $0 < E_L, E_R < \infty$, we can therefore expect *at least* three things: 4-component Dirac neutrinos, a vacuum filled with negative energy and two kinds of stable particles of non-zero rest mass.

The analysis presented here is in the framework of quantum mechanics of particles. The next step would be to consider such a model in the framework of quantum field theory.

References

[1] R. D. McKewon and P. Vogel, *Phys. Reports* **394** (2004) 315–356.

[2] W. Heisenberg, *Rev. Mod. Phys.* **29** (1957) 269.

[3] W. Heisenberg *et al.*, *Zeit. f. Naturforschung* **14a** (1959) 441.

[4] D. K. Sen, *Nuovo Cimento* **31** (1964) 660.

[5] D. K. Sen, *J. Math. Phys.* **48** (2007) 022304 1-8.

[6] L. de Broglie, *Compt. Rend.* **185** (1932) 536; **187** (1932) 1377.

[7] P. Jordan, *Zeits. Phys.* **93** (1935) 464.

[8] M. H. L. Pryce, *Proc. Roy. Soc. A.* **165** (1935) 247.

[9] F. M. Barbour, A. Bietti and B. F. Touschek, *Nuovo Cimento* **28** (1963) 452.

[10] V. V. Dvoeglazov, *Annales Fond. L. de Broglie* **24** (1999) 111.

[11] Y. Choquet-Bruhat, C. DeWitt and M. Dillard-Bleick, *Analysis, Manifolds and Physics* (North-Holland, Amsterdam, New York, 1978).

[12] D. K. Sen, *Fields and/or Particles* (Ryerson and Academic Press, Toronto, New York, 1968).

[13] A. Bohm, *Quantum Mechanics* (Springer, New York, 2001).

[14] F. Schwabl, *Quantum Mechanics* (Springer, Berlin, Heidelberg, 2002).

APPENDIX A

VECTOR FIELDS ON MANIFOLDS

A.1. Vector Fields on Manifolds

M will always denote an n-dimensional manifold, $T_m(M)$ the tangent space at $m \in M$, $X_m \in T_m(M)$ a tangent vector at m. $T_m^*(M)$ will denote the cotangent space of co-vectors w_m at m. The sets $TM = \bigcup_{m \in M} T_m(M)$ and $T^*M = \bigcup_{m \in M} T_m^*(M)$ are called the tangent and cotangent bundles of M, respectively, and they have natural manifold structures. Let p_M, q_M denote the corresponding projection maps, $p_M(X_m) = m = q_M(w_m)$:

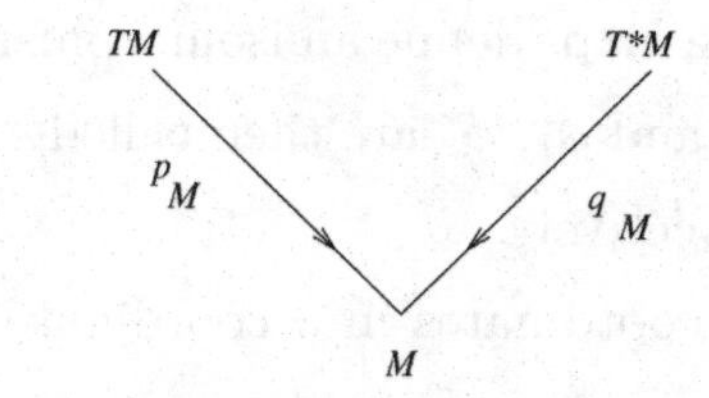

A vector field X is a cross-section of TM, i.e., a map $X : M \to TM$ such that $p_M \circ X = Id_M$. Similarly, a covector field or 1-form w is a cross-section of T^*M, i.e., a map $w : M \to T^*M$ such that $q_M \circ w = Id_M$. A tensor field of type (r, s) is similarly a cross-section of the tensor bundle

$$\bigcup_{m\in M} (\otimes T_m(M))^r \ \otimes ((\otimes T_m^*(M))^s)$$

and an exterior p-form ($p \geq 1$) is a cross-section of the bundle of *exterior p forms*: $\bigcup_{m\in M} \Lambda p \ T_m(M)$. A 0-form $f \in C^\infty(M)$ is a smooth function on M. Tensor fields and exterior forms on M form tensor algebra $T(M)$ and exterior algebra $\Lambda(M)$ over M.

If $\phi : M \to N$ is a map from M into a manifold N, we shall denote the differential of ϕ by ϕ_*, which can be regarded as a map $\phi_* : TM \to TN$, and the co-differential of ϕ by ϕ^*, which is a map $\phi^* : T^*N \to M$. If we are interested at a point $m \in M$, then ϕ_{*m} is a map $\phi_{*m} : T_m(M) \to T_{\phi(m)}(N)$ and $\phi^*_m : T^*_{\phi(m)}(N) \to T^*_m(M)$. One can extend these maps to the tensor products $(\phi T_m(M))^r$ and $(\phi T^*_\phi(m)(N))^s$, respectively. While ϕ_*, in general, does not take a vector field on M into vector fields on N, ϕ^* does take forms on N into forms on M, and we shall use the same symbol ϕ^* for the "pull-back" of forms on N into forms on M. If, however, $\phi : M \to N$ is a diffeomorphism, then these maps define an isomorphism of the tensor algebras $\tau(M)$ and $\tau(N)$, and ϕ_*, ϕ^* are then called "push-forward" and "pull-back" of ϕ, respectively.

If $(q_1, \ldots, q_n)$ are local coordinates in a coordinate neighborhood $U \subset M$, then $\frac{\partial}{\partial q_i}$ form a basis for $T_q(M)$, so that a tangent vector $v \in T_q(M)$, $q \in T_q(M)$, can be written as $v = \sum v_i \frac{\partial}{\partial q_i}$. Thus, $(q_1, \ldots, q_n, v_1, \ldots, v_n)$ form a local coordinate system on $p_M^{-1}(U) \subset TM$ of the tangent bundle. Similarly, a co-vector p at q can be written in the dual basis $dq_{i q\in U}$ as $p = \sum p_i dq_i$. Hence, $(q_1, \ldots, q_n, p_1, \ldots, p_n)$ form a local coordinate system on $q_M^{-1}(U) \subset T^*M$ of the cotangent bundle. A vector field X on M can be expressed in the local coordinate system as[f] $X = \sum X_i(q)\frac{\partial}{\partial q_i}$ which is equivalent to $X : (q) \to (q, X_i)$ or $(q_1, \ldots, q_n) \to (q_1, \ldots, q_n, X_1, \ldots, X_n)$.

Now let X be a vector field on the tangent bundle TM, i.e., a map $X : TM \to TTM$, such that and $p_{TM} \circ X = Id_{TM}$. Here TTM is the second tangent bundle and p_{TM} the natural projection on TM. In the above system of local coordinates for TM, we can express X as

$$X = \sum a_i(q, v)\frac{\partial}{\partial v_i} + \sum b_i(q, v)\frac{\partial}{\partial v_i},$$

which is equivalent to $X : TM \to TTM$ given by $X : (q, v) \to (q, v, a, b)$.

An *integral* curve of a vector field X on M is a map $\gamma : I \to M$, from an interval in $\mathbb{R}$ into M, such that $X \circ \gamma = \gamma * |_{Ix\{1\}} : I \to TM$. In local coordinates this means

$$\frac{dq_i}{dt} = X_i(q)$$

where $t \mapsto q(t)$ is the local expression for $\gamma : I \to M$.

Conversely, given a point $m \in M$ and a curve $\gamma : I \to M$ passing through m there exists a neighborhood $N(m)$ and a (local) vector field X on $N(m)$ such that γ is an integral curve of X. For a proof, see [4] and for an example, see below.

A vector field X on M generates a local 1-parameter group of local diffeomorphisms on M, that is, for every point $m \in M$, a map $\chi : U \times I \to M$, where U is a neighborhood of m and I an open

---

[f] We are *abusing* our notations here a bit by *confounding* a map by its value at a point and a point by its coordinate representative, or a vector by its components.

interval of R containing the origin, such that

1. for every $t \in I$, $\chi_t : U \to \chi_t(U)$ is a diffeomorphism,
2. $\{\chi_t\}$ form a *local* group on I with $\chi_0 = Id$ and $\chi_s \circ \chi_t = \chi_{t+s}$,

where χ_t is called a local *flow* of X.

The exterior derivative operator $d : \wedge(M) \to \wedge(M)$, taking p-forms into $p+1$-forms and the interior product operator $i_X : \wedge(M) \to \wedge(M)$ taking p-forms into $p-1$-forms, are related to the Lie-derivative $L_X : \wedge(M) \to \wedge(M)$ by means of the basic identity $i_X d + d i_X = L_X$. Also, $i_{[X,Y]} = [L_X, i_Y]$. The Lie-derivative, of course, can be also defined on $\tau(M)$, as follows. If $K \in \tau(M)$, define for every diffeomorphism $\phi : M \to M$, $\tilde{\phi}K \in \tau(M)$ by $(\tilde{\phi}K)(m) = \phi^*_m(K(\phi(m)))$, $m \in M$, i.e., we pull back $K(\phi(m))$ to m. Let X generate the local flow χ_t, through its integral curves, then[g]

$$L_X K = \lim_{t \to O} \frac{1}{t}[\tilde{\chi}_t K - K].$$

The Lie-derivative is natural with respect to the push-forward map $\phi_* : \tau(M) \to \tau(N)$ where $\phi : M \to N$ is a diffeomorphism, i.e., $\phi_*(L_X K) = L_{\phi_* X}(\phi_* K)$. Hence, $\tilde{\chi}_t(L_X K) = L_X(\tilde{\chi}_t K)$, since $\tilde{\chi}_t(L_X K) = \frac{d}{dt}(\tilde{\chi}_t K)$. Also, from the definition of Lie-derivative, a tensor field K is constant along the integral curves of X if $L_X K = 0$.

A p-form α is *closed* if $d\alpha = 0$. It is *exact* if $\alpha = d\beta$, where β is a $p - l$-form. An exact form is thus closed. The converse, known as the Poincaré Lemma, is true only locally, i.e., if α is closed, then for every $m \in zM$, there is a neighborhood U of m, on which α is exact.

---

[g]Our definition of $\tilde{\phi}$ is somewhat different from that of [5]. Hence, the difference in sign.

A.2. Example of a Vector Field whose Integral Curves are Knots

The following is an example of a local vector field on R^3 whose integral curves are *torus knots*, i.e., knots which lie on a 2-torus T^2 embedded in the Euclidean space R^3 ([5]).

Consider the map from $[0, 2\pi]$ into R^3

$$
\begin{aligned}
&t \mapsto (u_1 = \alpha_1 t, u_2 = \alpha_2 t)\\
&(u_1, u_2) \mapsto x(u_1, u_2)\\
&\quad = ((b + a\cos(u_1))\cos(u_2), (b + a\cos(u_1))\sin(u_2), a\sin(u_1)).
\end{aligned} \tag{A.1}
$$

Here, α_1, α_2 are *relatively prime* integers, that is, $(\alpha_1, \alpha_2) = 1$; a, b are the inner and outer radii and u_1, u_2 the inner and outer angular parameters, respectively, of T^2. The above map describes a knot which winds α_1-times around the inner circle and α_2-times around the outer circle of the torus.

Since $\frac{du_1}{dt} = \alpha_1$, $\frac{du_2}{dt} = \alpha_2$, the vector field on T^2 whose integral curves are torus knots is

$$
X = \sum_i \alpha_i \frac{\partial}{\partial u_i} = \alpha_1 \frac{\partial}{\partial u_1} + \alpha_2 \frac{\partial}{\partial u_2}. \tag{A.2}
$$

We now extend *locally* the vector field X on T^2 to a vector field $\tilde{X}$ on R^3 by the embedding of T^2 in R^3 as follows:

$$
\begin{aligned}
\tilde{X} &= \sum_k \beta_k(x) \frac{\partial}{\partial x_k},\\
\beta_k(x) &= \sum_i \alpha_i \frac{\partial x_k}{\partial u_i}.
\end{aligned} \tag{A.3}
$$

Now,

$$\begin{aligned} \beta_1 &= -\alpha_1 x_3 \cos(u_2) - \alpha_2 x_2, \\ \beta_2 &= -\alpha_1 x_3 \sin(u_2) + \alpha_2 x_1, \\ \beta_3 &= \alpha_1 a \cos(u_1). \end{aligned} \tag{A.4}$$

Since on T^2

$$\begin{aligned} \tan(u_2) &= \frac{x_2}{x_1}, \\ a\cos(u_1) &= \frac{x_1}{\cos(u_2)} - b = \frac{x_2}{\sin(u_2)} - b, \end{aligned} \tag{A.5}$$

we can express β_k in terms of x_i to get finally

$$\begin{aligned} \tilde{X} = &\left(-\frac{\alpha_1 x_3 x_1}{\sqrt{x_1^2 + x_2^2}} - \alpha_2 x_2\right)\frac{\partial}{\partial x_1} \\ &+ \left(-\frac{\alpha_1 x_3 x_2}{\sqrt{x_1^2 + x_2^2}} + \alpha_2 x_1\right)\frac{\partial}{\partial x_2} + \alpha_1\left(\sqrt{x_1^2 + x_2^2} - b\right)\frac{\partial}{\partial x_3}. \end{aligned} \tag{A.6}$$

The integral curves of $\tilde{X}$ are, therefore, given by the differential equations

$$\begin{aligned} \frac{dx_1}{dt} &= -\frac{\alpha_1 x_3 x_1}{\sqrt{x_1^2 + x_2^2}} - \alpha_2 x_2, \\ \frac{dx_2}{dt} &= -\frac{\alpha_1 x_3 x_2}{\sqrt{x_1^2 + x_2^2}} + \alpha_2 x_1, \\ \frac{dx_3}{dt} &= \alpha_1\left(\sqrt{x_1^2 + x_2^2} - b\right). \end{aligned} \tag{A.7}$$

Figure 5 shows a torus knot obtained by numerical integration of the above equations and the corresponding torus for $\alpha_1 = 3$, $\alpha_2 = 2$, $a = 1, b = 2$.

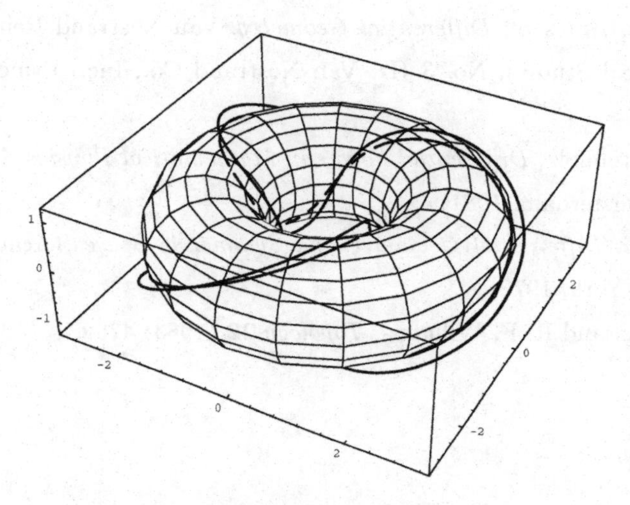

Fig. 5. Torus knot.

References

[1] S. Kobayashi and K. Nomizu, *Foundations of Differential Geometry*, Vol. 1 (Wiley, New York, 1963).

[2] N. J. Hicks, *Notes on Differential Geometry*, Van Nostrand Reinhold Mathematical Studies, No. 3 (D. Van Nostrand Co., Inc., Princeton, 1965).

[3] C. von Westenholz, *Differential Forms in Mathematical Physics* (North Holland, Amsterdam, 1981).

[4] S. Helgason, *Differential Geometry and Symmetric Spaces* (Academic Press, New York, 1972).

[5] J. S. Birman and R. F. Williams, *Topology* **22** (1983) 4782.

APPENDIX B

DYNAMICAL VECTOR FIELDS OF CLASSICAL MECHANICS

B.1. Introduction

Dynamical vector fields are vector fields which describe the motion of a system of particles in classical mechanics. In the course of history, classical mechanics has evolved into various different formalisms — Newtonian, Lagrangian, Hamiltonian, Hamilton–Jacobi, etc. Each of these formalisms can be described geometrically on manifolds[h] with corresponding appropriate geometric structures — second-order ordinary differential equations (O.D.E.), sprays, symplectic and contact structures, etc. We first summarize briefly the salient features of each of the classical formalisms ([1, 2]). This will motivate the consideration of these formalisms later in a somewhat abstract differential geometric setup.

Newtonian Formalism. We shall illustrate the Newtonian formalism by considering the example of a system of N particles moving in $\mathbb{R}^3$. Each particle traces out a curve in $\mathbb{R}^3$, i.e., the graph of a map $\vec{x}_\alpha : \mathbb{R} \to \mathbb{R}^3$, $(a = 1, \ldots, N)$, where $\mathbb{N}$ is the time axis. The motion

[h]By manifolds we shall always mean smooth manifolds. Also all maps will be smooth unless specified otherwise.

of such a system is then described by the Newton's equations

$$m_\alpha \ddot{\vec{x}}_\alpha = \vec{F}_\alpha, \tag{B.1}$$

where $m_\alpha > 0$ are the masses and $\vec{F}_\alpha$, the force functions. Let $\vec{x}_\alpha = (x_\alpha, y_\alpha, z_\alpha)$, $\vec{F}_\alpha = (F_{\alpha x}, F_{\alpha y}, F_{\alpha z})$ and set

$$\begin{cases} x_1 = q_1, \quad y_1 = q_2, \quad z_1 = q_3; \quad x_2 = q_4, \quad y_2 = q_5, \quad z_2 = q_6, \text{ etc.} \\ M_1 = M_2 = M_3 = m_1, \quad M_4 = M_5 = M_6 = m_2, \text{ etc.} \\ F_1 = F_{1x}, \quad F_2 = F_{1y}, \quad F_3 = F_{1z}, \text{ etc.} \end{cases}$$

So that (B.1) can be written as

$$\ddot{q}_i = \frac{F_i}{M_i}(q_1, \ldots, q_n, \dot{q}_1, \ldots, \dot{q}_n) \quad (i = 1, \ldots, n = 3N) \tag{B.2}$$

where we have excluded the case of time-dependent forces and included the fact that the force functions, in general, depend on the velocities as well as positions. Note that each term of the right-hand side of (B.2) is (in general) the value of a map: $\mathbb{R}^n \times \mathbb{R}^n \to \mathbb{R}$.

B.2. Configuration Spaces of Mechanical Systems as Manifolds

Associated with each mechanical system is the so-called *configuration space*, which is roughly the space available to the system to move about. The dimension of the configuration space is the number of degrees of freedom of the system. In the above example of N particles in $\mathbb{R}^3$, if there are no *constraints*, the configuration space is $\mathbb{R}^n$ which is, of course, a manifold. However, if we wish to exclude the *collision* points, that is, those points of $\mathbb{R}^{3N}$, for which, $\vec{x}_\alpha = \vec{x}_\beta$, $\alpha \neq \beta$, the configuration space is then an *open* submanifold of $\mathbb{R}^{3N}$.

If we now consider a system of N particles in $\mathbb{R}^3$ subject to k *holonomic* constraints given by:

$$f_a(x_1, \ldots, x_N, Y_1, \ldots, Y_N, z_1, \ldots, z_N) = 0, \quad a = 1, \ldots, k \tag{B.3}$$

we can choose a new set of coordinates $u_1, \ldots, u_{3N}$ as follows

$$\begin{aligned} u_1 &= u_1(x_1, \ldots, x_N, \ldots) \\ &\vdots \\ u_{3N-k} &= u_{3N-k}(x_1, \ldots, x_N, \ldots) \\ u_{3N-k+k} &= f_1(x_1, \ldots, x_N, \ldots) \\ &\vdots \\ u_{3N} &= f_k(x_1, \ldots, x_N, \ldots) \end{aligned}$$

so that $u_{3N-k+k} = \cdots = u_{3N} = 0$, in view of (B.3). Thus, the system can be described on a *submanifold* of $\mathbb{R}^{3N}$ of dimension $n = 3N - k$. This follows from the so-called *submanifold property.* Recall that a manifold M is a submanifold of a manifold M' iff (i) $M \subset M'$ as top subspace and (ii) the inclusion map $i : M \hookrightarrow M'$ is an *embedding*, i.e., $i_{*_p} : T_p(M) \to T_{i(p)}(M')$ is injective.

Proposition (Submanifold Property). *There exists a coordinate system* $(x_1, \ldots, x_m)$ *valid on an open neighborhood* V *of every point* $p \in M \subset M'$ $(\dim M' = m \geq n = \dim M)$ *in* M' *such that*

(i) $x_1(p) = \cdots = x_m(p) = 0.$

(ii) $U = \{q \in V | x_j(q) = 0$ *for* $n+1 \leq j \leq m\}$, *together with the restriction of* $(x_1, \ldots, x_n)$ *to* U *form a local chart of* p *in* M.

Note. When $\dim M = \dim M', M$ is an open submanifold of M'. The configuration space of a system of N particles with holonomic[i] constraints is, thus, a $(n = 3N - k)$-dimensional manifold.

Other examples of such configuration spaces are:

Mechanical System	Configuration Space
Planar pendulum	S^1
Sperical pendulum	S^2
Planar double pendulum	$S^1 \times S^1 = T^2$
Rigid rod in space	$S^2 \times \mathbb{R}^3$
Rigid body free to rotate about a point	$SO(3, \mathbb{R})$
Rigid body in space	$SO(3, \mathbb{R}) \times \mathbb{R}^3$

[i]We shall have no occasions to consider non-holonomic systems.

Problem. Show that $SO(3, \mathbb{R}) \simeq \mathbb{R}P^3$ (3-dimensional real projective space).

We are thus led to study mechanical systems on abstract manifolds.

Lagrangian Formalism. We illustrate the Lagrangian formalism by considering a mechanical system which is holonomic and conservative. Let $q = (q_1, \ldots, q_n)$ now be any "generalized" coordinate on the configuration space. The Lagrangian L is a function of q and the "generalized" velocities $\dot{q}$, i.e., a map $L : U \times V \to \mathbb{R}$, $U, V \subset \mathbb{R}^n$, defined on the "position-velocity" space by

$$L(q, \dot{q}) = T(\dot{q}) - V(q),$$

where T and V are the kinetic and potential energies, respectively. The Lagrangian equations of motion describing the evolution of the system are the Euler–Lagrange equations

$$\frac{\partial L}{\partial q_i} - \frac{d}{dt}\left(\frac{\partial L}{\partial \dot{q}_1}\right) = 0 \tag{B.4}$$

of the variational principle

$$\delta \int_{t_1}^{t_2} L(q, \dot{q})dt = 0, \tag{B.5}$$

with $$\delta q(t_1) = \delta q(t_2) = 0.$$

Legendre Transformations ([3]). The passage from the Lagrangian to the Hamiltonian formalism is realized by means of Legendre transformations which, mathematically, transform functions on a vector space (tangent bundle) to functions on its dual vector space (cotangent bundle).

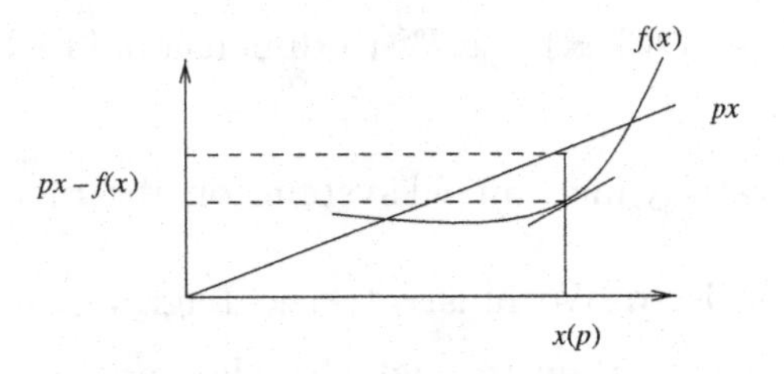

Fig. 6. Legendre transformation.

Let f be a smooth convex function of a single variable x (i.e., $f''(x) > 0$). We define a new function g of a new variable p as follows: For every number p, consider the difference $F(p,x) = px - f(x)$. It has a maximum with respect to x (see Fig. 6), given by $\frac{\partial F}{\partial x} = 0$.

Then, $g(p) = F(p, x(p))$. In other words, $g(p)$ is obtained by expressing x as a function of p by means of the equation $p = f'(x)$ and substituting in $F(p,x)$. Convexity of f assures that $x(p)$ is unique.

Example.

$$f(x) = m\frac{x^2}{2}, \quad F(p,x) = px - m\frac{x^2}{2},$$

so

$$x(p) = \frac{p}{m}, \quad g(p) = \frac{p^2}{2m},$$

$$\frac{\partial F}{\partial x} = p - mx = 0 \Rightarrow x = p/m = x(p),$$

$$g(p) = F(p, x(p)) = p \cdot \frac{p}{m} - \frac{m}{2}\left(\frac{p}{m}\right)^2 = \frac{p^2}{2m}.$$

Note.

$$f : \mathbb{R} \to \mathbb{R}, \quad g : \mathbb{R}^* \to \mathbb{R}^* \to \mathbb{R} \quad \text{where } \mathbb{R}^* \ni p : \mathbb{R} \to \mathbb{R},$$

$$x \mapsto f(x), \quad p \mapsto g(p) \qquad x \mapsto p(x) = \langle x, p \rangle = px.$$

A Legendre transformation takes a convex function into another convex function, so that one can apply the transformation again.

Proposition. *The Legendre transformation is involutive, i.e., the product of two successive transformations gives the identity.*

Proof. We start with a function f of the variable x and consider $F(p,x) = px - f(x)$. Then, $\frac{\partial F}{\partial x} = 0 \Leftrightarrow p = f'(x)$. Since $f''(x) \neq 0$, we can write $x = f'^{-1}(p)$ and, thus, $g(p) = F(p, f'^{-1}(p)) = pf'^{-1}(p) - f(f'^{-1}(p))$. We now consider another transformation to a new variable which we label again as x and so, let $G(x,p) = xp - g(p)$. So that $\frac{\partial G}{\partial p} = 0 \Leftrightarrow x = g'(p)$ or $p = g'^{-1}(x)$. Consider now $G(x, g'^{-1}(x)) = xg'^{-1}(x) - g(g'^{-1}(x))$. We want to show that this is equal to $f(x)$. Note that from the relationship between the two variables, $f'^{-1} \circ g'^{-1} = Id$. And, from the expression for $g(p)$, we have

$$\begin{aligned} g(g'^{-1}(x)) &= g'^{-1}(x) f'^{-1}(g'^{-1}(x)) - f(f'^{-1}(g'^{-1}(x))) \\ &= xg'^{-1}(x) - f(x). \end{aligned}$$

So $G(x, g'^{-1}(x)) = f(x)$. □

We can now generalize to the many variable case. Let $x = (x_1, \ldots, x_n)$, and f a smooth convex function of x, i.e., $\frac{\partial^2 f}{\partial x_i \partial x_k}$ is positive definite. Introduce a new function g of a new variable $p = (p_1, \ldots, p_n)$ as follows. Let $F(p,x) = \sum p_i x_i - f(x)$. The maximum of $F(p,x)$ with respect to x is obtained by setting $\frac{\partial F}{\partial x_i} = 0 \Leftrightarrow p_i = \frac{\partial f}{\partial x_i}$. Then $g(p) = F(p, x(p))$.

Example. $f : \mathbb{R}^n \to \mathbb{R},\ f(x) = \sum a_{ij} x_i x_j$, a_{ij} positive definite.

$$\frac{\partial F}{\partial x_i} = 0 \Leftrightarrow p_i = 2 \sum a_{ij} x_j,$$

that is, $p = 2ax$ as matrices, or $x = ca$, where $c = \frac{a^{-1}}{2}p$, i.e., $x_i = \sum_j c_{ij}p_j$. Then,

$$g(p) = F(p, x(p)) = 2\sum a_{ij}x_ix_j - \sum a_{ij}x_ix_j$$
$$= \sum b_{ij}p_ip_j$$

where b_{ij} is also positive definite.

Note that $g : \mathbb{R}^{n*} \to \mathbb{R}$, $p \in \mathbb{R}^{n*}$, $\langle p, x\rangle = \sum p_ix_i$.

Hamilton Formalism. Let us assume that the Lagrangian $L(q, \dot{q})$ is convex with respect to the variable $\dot{q} = (\dot{q}_1, \ldots, \dot{q}_n)$, i.e., $\frac{\partial^2 L}{\partial\dot{q}_i\partial\dot{q}_k}$ is positive definite. Now consider a Legendre transformation from the variable $\dot{q}$ to the new variable $p = (p_1, \ldots, p_n)$ regarding q as a *parameter*. Thus, we form the function

$$F(p, \dot{q}, q) = \sum p_i\dot{q}_i - L(q, \dot{q}) \tag{B.6}$$

and look for maximum of F

$$H(p, q) = F(p, \dot{q}(p, q), q) = \sum p_i\dot{q}_i(p, q) - L(q, \dot{q}(p, q)), \tag{B.7}$$

where H, the Hamiltonian, is a map $H : U \times W \to \mathbb{R}^*$, $W \subset \mathbb{R}^{n*}$, defined on the "position-momentum" space, i.e., the phase space. To obtain the Hamiltonian equations of motion, one simply takes the total differential of both sides of (B.7) and use (B.4) and (B.6)

$$\sum \frac{\partial H}{\partial q_i}dq_i + \sum \frac{\partial H}{\partial p_i}dp_i = \sum p_1 d\dot{q}_i + \sum \dot{q}_i dp_i - \sum \frac{\partial L}{\partial q_i}dq_i,$$
$$-\sum \frac{\partial L}{\partial \dot{q}_i}d\dot{q}_i = -\sum \dot{p}_i dq_i + \sum \dot{q}_i dp_i$$

so that

$$\left.\begin{aligned} \dot{q}_i &= \frac{\partial H}{\partial p_i} \\ \dot{p}_i &= -\frac{\partial H}{\partial q_i}. \end{aligned}\right\} \tag{B.8}$$

One can rewrite (B.8) using Poisson brackets as follows. Let $f : U \times W \to \mathbb{R}$, $(q,p) \mapsto f(q,p)$ be any "observable" on the phase space. Then

$$\dot{f} = \sum \frac{\partial f}{\partial q_i}\dot{q}_i + \sum \frac{\partial f}{\partial p_i}\dot{p}_i = \sum \left(\frac{\partial f}{\partial q_i}\frac{\partial H}{\partial p_i} - \frac{\partial f}{\partial p_i}\frac{\partial H}{\partial q_i}\right). \tag{B.9}$$

The right-hand side of (B.9) is the Poisson bracket of f and H and is denoted by $\{f, H\}$. Thus (B.8) can be written in the P.B. form as

$$\left.\begin{aligned} \dot{q}_i &= \{q_i, H\} \\ \dot{p}_i &= \{p_i, H\}. \end{aligned}\right\} \tag{B.10}$$

A *first integral* of a Hamiltonian system is an observable f whose time derivative vanishes. This means that $\{f, H\} = 0$. Note that since $\{H, H\} = 0$, H itself is a first integral.

It is clear from the involutive property of Legendre transformations that a second Legendre transformation takes the Hamiltonian back to the Lagrangian.

Canonical Transformations and Symplectic Structure. Let us rewrite the Hamiltonian equations (B.8) in a slightly different form by setting $\xi = (q,p) = (q_1, \ldots, q_n, p_1, \ldots, p_n)$. Then $\mathrm{grad}_\xi H = (\frac{\partial H}{\partial q_1}, \ldots, \frac{\partial H}{\partial q_n}, \frac{\partial H}{\partial p_1}, \ldots, \frac{\partial H}{\partial p_n})$ with $H = H(\xi)$. Equation (B.8) can now be written as

$$\left.\begin{aligned} &\dot{\xi} = J \cdot \mathrm{grad}_\xi H \\ &\text{where} \quad J = \begin{pmatrix} 0 & I \\ -I & 0 \end{pmatrix}, \quad I = n \times n \quad \text{unit matrix.} \end{aligned}\right\} \tag{B.11}$$

Consider now a transformation to a new set of variables $n = (\bar{q}, \bar{p}) = f(\xi)$ with non-singular Jacobian $A = (\frac{\partial n_\alpha}{\partial \xi_\beta})$, $\alpha, \beta = 1, \ldots, 2n$.

Then,

$$\dot{\eta} = A\dot{\xi} = AJ \text{ grad}_\xi H(\xi) = AJA^T \text{ grad}_\eta H(f^{-1}(\eta)).$$

The transformation $\xi \mapsto \eta$ will preserve the form of the Hamiltonian equations (B.11) iff

$$AJA^T = J. \tag{B.12}$$

Such a transformation is called *canonical* or symplectic.

Problems.

(1) Show that the canonical transformations form a group.

(2) Let $\bar{q}_i = \bar{q}_i(q, p)$, $\bar{p}_i = \bar{p}_i(q, p)$ be a canonical transformation. Then, show that

$$\frac{\partial \bar{p}_i}{\partial q_k} = -\frac{\partial p_k}{\partial \bar{q}_i} \quad \text{and} \quad \frac{\partial \bar{p}_i}{\partial p_k} = \frac{\partial q_k}{\partial \bar{q}_i}.$$

(3) Show that the following are canonical transformations:

(i) $\bar{q}_i = \bar{q}_i(q)$, $A = (\frac{\partial \bar{q}_i}{\partial q_j})$ non-singular.

(ii) $p_i = \sum (\frac{\partial \bar{q}_j}{\partial q_i}) \bar{p}_j$ (point transformation).

(iii) $n = 1$, $\bar{q} = \alpha q + \beta p$, $\bar{p} = \gamma q + \delta p$ with $\alpha\delta - \beta\gamma = 1$.

(iv) $\bar{q}_i = p_i$, $\bar{p}_i = -q_i$.

A *linear* transformation $\mathbb{R}^{2n} \to \mathbb{R}^{2n}$ is said to be symplectic if the corresponding transformation (constant) matrix A satisfies (B.12). Such a matrix A, known as a *symplectic matrix*, preserves the so-called (skew symmetric bilinear) *symplectic form* $w : \mathbb{R}^{2n} \times \mathbb{R}^{2n} \to \mathbb{R}$ given by $w(v_1, v_2) = \langle v_1, Jv_2 \rangle$. In order to understand the Hamiltonian formalism, we are thus led to the study of symplectic geometry and symplectic manifolds.

Hamilton–Jacobi Formalism. We now consider time-dependent systems where the Lagrangian L as well as the Hamiltonian H depend *explicitly* on time t. The (q, p, t) space is known as the *extended phase space.* The relationship between L and H is still given by (B.7), i.e., $H(q, p, t) = \sum p_i \dot{q}_i - L(q, \dot{q}, t)$ and the equations of motion still follow from the variational principle (B.5), which written in terms of H becomes

$$\delta \int_1^2 \left(\sum p_i \dot{q}_i - H \right) dt = 0. \tag{B.13a}$$

Let us now consider a transformation of the extended phase space $(q, p, t) \mapsto (\bar{q}, \bar{p}, t)$, with $\bar{q}_i = \bar{q}_i(q, p, t)$, $\bar{p}_i = \bar{p}_i(q, p, t)$, such that the new equations of motion are again derived from a corresponding variational principle

$$\delta \int_1^2 \left(\sum \bar{p}_i \dot{\bar{q}}_i - \overline{H} \right) dt = 0. \tag{B.13b}$$

Such transformations are known as *time-dependent canonical* transformations or sometimes *contact* transformations.[j] Equations (B.13a) and (B.13b) show that the integrands in (B.13a) and (B.13b) differ *locally* by the total time derivative of a function F

$$\sum p_i \dot{q}_i - H - \left(\sum \dot{p}_i \dot{\bar{q}}_i - \overline{H} \right) = \frac{dF}{dt}. \tag{B.14}$$

Now F depends on the new as well as the old variables $q_i, p_i, \bar{q}_i, \bar{p}_i$, and t. However, since they are related by $2n$ equations, only $2n + 1$ of them constitute an independent set of variables. Thus, we have

[j]There is a considerable variation in the literature about the use of the terms *canonical* and *contact* transformations. The time-independent canonical transformations are also sometimes called *homogeneous contact* transformations.

the following choice for F:

$$F_1(q, \bar{q}, t), \quad F_2(q, \bar{p}, t), \quad F_3(p, \bar{q}, t), \quad F_4(p, \bar{p}, t).$$

Such functions are called *generating functions* of the contact transformations. Consider a transformation generated by $F_1(q, \bar{q}, t)$. We have from (B.14)

$$\sum p_i \dot{q}_i - \sum \bar{p}_i \dot{\bar{q}}_i + \overline{H} - H = \sum \frac{\partial F_1}{\partial q_i} \dot{q}_i + \sum \frac{\partial F_1}{\partial \bar{q}_i} \dot{\bar{q}}_i + \frac{\partial F_1}{\partial t}$$

or

$$p_i = \frac{\partial F_1}{\partial q_i}, \quad \bar{p}_i = \frac{\partial F_1}{\partial \bar{q}_i}, \quad \overline{H} = H + \frac{\partial F_1}{\partial t}. \tag{B.15}$$

We say that the contact transformation transforms the Hamiltonian to *equilibrium* if $\overline{H}$ = const., i.e., $\frac{\partial \overline{H}}{\partial \bar{q}_i} = \frac{\partial \overline{H}}{\partial \bar{p}_i} = \frac{\partial \overline{H}}{\partial t} = 0$. Since a contact transformation preserves the form of Hamiltonian equations, $\dot{\bar{q}}_i = \frac{\partial \overline{H}}{\partial \bar{p}_i} = 0$ and $\dot{\bar{p}}_i = -\frac{\partial \overline{H}}{\partial \bar{q}_i} = 0$. Thus, $\bar{q} = a = (a_1, \ldots, a_n) =$ const. Setting $F_1(q, \bar{q}, t) \equiv S(q, a, t)$, the last equation of (B.15) becomes

$$H\left(q_1, \ldots, q_n, \; p_1 = \frac{\partial S}{\partial q_1}, \ldots, p_n = \frac{\partial S}{\partial q_n}, t\right) + \frac{\partial S}{\partial t} = \text{const.}, \tag{B.16}$$

which is the *time-dependent Hamilton–Jacobi equation.* The equilibrium generating function S, known as *Hamilton's principal function,* has another significance, besides others. We have $\frac{dS}{dt} = \sum \frac{\partial S}{\partial q_i} \dot{q}_i + \frac{\partial S}{\partial t} = \sum p_i \dot{q}_i - H + \text{const.} = L + \text{const.}$ The constant in (B.16) can be taken to be zero by letting $S \mapsto S - (\text{const.})t$, so that finally

$$S = \int L \, dt + \text{const.} \tag{B.17}$$

S is therefore also called the *action* function.

B.3. Dynamical Vector Fields in Classical Mechanics

The Geometrical Picture ([4]). Our brief summary of the formalisms of classical mechanics has revealed to us its two essential features — firstly, that the configuration space is, in general, a smooth manifold. Secondly, the equations of motion, for example, the Hamiltonian equations, exhibit a certain structure (namely, the symplectic structure).

In order to describe the temporal evolution of a mechanical system, it is necessary to go beyond the configuration space to the so-called "state-space" S, which is roughly the set (manifold) of all possible "states" of the system. For example, the state-space would be either the position-velocity space or the phase space, which will turn out to be the tangent bundle or the cotangent bundle, respectively, of the configuration space. If initially, at $t = 0$, the system is in the state given by $x \in S$, its state at any time $t > 0$ can be specified by a smooth map $\chi - t : S \to S$, $x \mapsto \chi_t(x)$, with the property that $\chi_{t_2} \circ \chi_{t_1} = \chi_{t_1+t_2}$. For reversible systems, $\chi_t^{-1} = \chi_t$ should exist $\forall\, t \in R$. In other words, the dynamics of the system can be described by a (global) 1-parameter group of (global) diffeomorphisms of S. Now, a global 1-parameter group of global diffeomorphisms of S determines a vector field X on S by assigning to $\forall\, x \in S$, the tangent vector to the curve $t \mapsto \chi_t(x)$ at $x = \chi_0(x)$. Conversely, the dynamics of the system can also be described by specifying the dynamical vector field X. However, then X generates, in general, only a *local* 1-parameter group of local diffeomorphisms, its (local) *flow*, unless of course X is *complete*. Our geometrical picture will then very frequently be a state-space S, a dynamical vector field X on S, whose flow describes the temporal evolution of the system, together with a certain flow-preserving structure of S.

Newton's Equations as a Second-Order O.D.E. ([6]). We can now look upon the Newtonian equations (B.2) from our stated geometrical viewpoint. Our state space S will be the position-velocity space, i.e., the tangent bundle TM of the configuration space M of a Newtonian system. We only have to find the dynamical vector field X on TM whose flows will describe the temporal evolution of the system according to Newton's equations (B.2).

Definition. A second-order ordinary differential equation (O.D.E.) on M is a vector field X on TM such that $(p_M)_* \circ X = Id_{TM}$.

Note that p_M is a map $p_M : TM \to M$. Hence, $(p_M)_*$ is a map $(p_M)_* : TTM \to TM$. Since X is a vector field on TM, we have $X : TM \to TTM$, where $p_{TM} \circ X = Id_{TM}$. In other words, the following diagram commutes

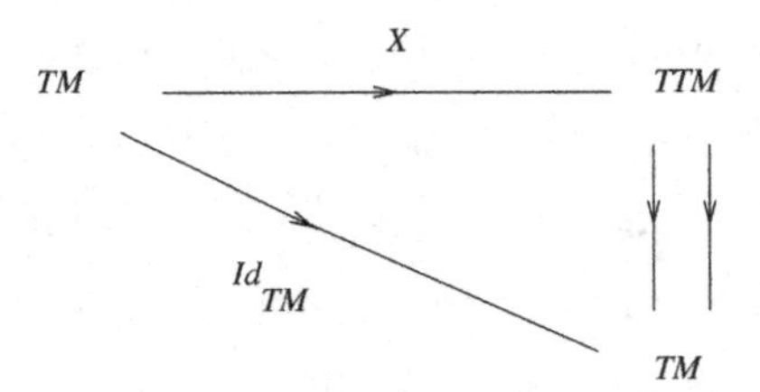

Definition. A solution of a second-order O.D.E. X on M is a curve $\gamma : I \to M$ such that $\gamma * |_{Ix\{1\}} : I \to TM$ is an integral curve of X on TM.

To see what the condition for X to be a second-order O.D.E. on M means, consider the local coordinate expression of an arbitrary vector field X on TM

$$X = \sum a_i(q, v)\frac{\partial}{\partial q_i} + \sum b_i(q, v)\frac{\partial}{\partial v_i}. \tag{B.18}$$

In local coordinates, $p_M : (q, v) \mapsto (q)$, and so

$$(p_M)_* \circ X : (q, v) \mapsto (q, v, a, b) \mapsto (q, a).$$

The condition $(p_M) \circ X = Id_{TM}$ means

$$a_i(q, v) = v_i \tag{B.19}$$

so that for a second-order O.D.E.,

$$X = \sum v_i \frac{\partial}{\partial q_i} + (q, v) \frac{\partial}{\partial v_i}. \tag{B.20}$$

The integral curves of X on TM are thus given by

$$\left.\begin{aligned} \frac{dq_i}{dt} &= v_i \\ \frac{dv_i}{dt} &= b_i(q, v). \end{aligned}\right] \tag{B.21}$$

That is, the solution curves of X on M satisfy

$$\frac{d^2 q_i}{dt^2} = b_i(q, \dot{q})$$

which are precisely the Newtonian equations (B.2) with $b_i = \frac{F_i}{M_i}$. Thus, the geometrical structure of the Newtonian formalism is (TM, X), where X is a second-order O.D.E. on TM.

A special type of second-order O.D.E. is a "*spray*", which has applications in mechanics. Define a vector field V on TM by its flow $\chi_t : (q, v) \mapsto (q, e^t v)$ in local coordinates. That is

$$V = \sum v_i \frac{\partial}{\partial v_i}. \tag{B.22}$$

V generates a l-parameter group of *homotheties* and is called *Liouville*.

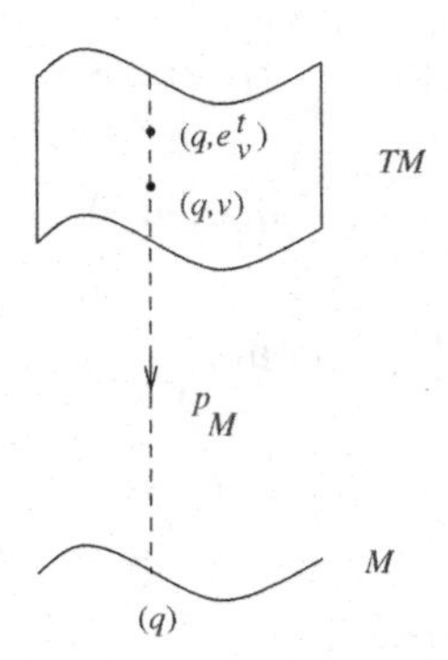

Definition. A spray on M is a second-order O.D.E. X on M such that $[V, X] = X$.

In local coordinates,

$$[V, X] = \sum v_i \frac{\partial}{\partial q_i} + \sum \left(v_i \frac{\partial}{\partial v_i} - b_j \right) \frac{\partial}{\partial v_j}$$

so

$$[V, X] = X \to \sum \left(v_i \frac{\partial b_j}{\partial v_i} = 2b_j \right). \tag{B.23}$$

That is, b_j are homogeneous functions of degree 2 in v_i. A geodesic spray of a connection ∇ on m is a spray whose solution curves are geodesics of ∇.

Symplectic Manifolds and Hamiltonian Systems ([7]). We now turn to the Hamiltonian equations (B.8) which, we have noted already, exhibit a symplectic structure.

A non-degenerate two-form w on a vector space V is a skew symmetric bilinear form $w : V \times V \to \mathbb{R}$, such that $w(v_1, v_2) = 0$ for $\forall v_2 \in V \Rightarrow v_1 = 0$. Let $\{e_a\}$ be a basis of V, $\{\alpha_a\}$ the dual basis of V^*, and $w(e_a, e_b) = w_{ab} = -w_b a$, the matrix of w relative to $\{e_a\}$. Non-degeneracy of w means w_{ab} is non-singular and this is possible only when the rank of $w_{ab} = 2n = \dim V$. Analogous to the symmetric case, one can then choose a basis such that

$$w = \sum_{i=1}^{n} \alpha_i \wedge \alpha_{i+n} \tag{B.24}$$

so that, $w_{ab} = \left(\begin{smallmatrix} 0 & I \\ -I & 0 \end{smallmatrix}\right) = J$, $I = n \times n$ identity matrix. A symplectic vector space is a pair (V, w), where w is a non-degenerate two-form on V. The set of all invertible linear maps $T : V \to V$ such that $w(v_1, v_2) = w(TV_1, TV_2)$ form a subgroup of $GL(V, V)$, called the *symplectic group* $Sp(V, w)$. In matrix form, the above condition means (denoting again by T its matrix) precisely the conditions for T to be a symplectic matrix, i.e.,

$$T^T JT = J. \tag{B.25}$$

Problem. Show that

(1) If $T = \left(\begin{smallmatrix} a & b \\ c & d \end{smallmatrix}\right)$, where a, b, c, d are $n \times n$ matrices, then T is symplectic $\Leftrightarrow a^T c$, $b^T d$ are symmetric and $a^T d - c^T b = I$.

(2) If $\lambda \in \mathbb{C}$ is an eigenvalue of T, then so are $\frac{1}{\lambda}, \bar{\lambda}, \frac{1}{\bar{\lambda}}$.

(3) The symplectic matrices form a Lie subgroup $Sp(2n, \mathbb{R}) \subset GL(2n, \mathbb{R})$ of dimension $2n^2 + n$.

Definition. (M, W) is a symplectic manifold iff

(i) M is a $2n$-dimensional manifold,

(ii) w is a non-degenerate exterior two-form on M, i.e., for each $m \in M$, $w(m)$ is a non-degenerate, skew-symmetric, two-form on $T_m(M)$,

(iii) $dw = 0$, i.e., w is closed.

Note that on a symplectic manifold (M, w), $w^n = w \wedge \cdots \wedge w$ is a volume-element, so that a symplectic manifold is orientable.

Darboux's Theorem ([7, 8]). *On a symplectic manifold (M, w), at every $m \in M$, there exists a coordinate neighborhood U with a coordinate system $(x_1, \ldots, x_n, y_1, \ldots, y_n)$, such that, on U*

$$w = \sum dx_i \wedge dy_i. \tag{B.26}$$

Proof. It is sufficient to assume that M is a linear space with m at its origin 0. Let w_1 be a constant form which equals $w(0)$ everywhere and $w_t = w + t(w_1 - w)$, $0 \leq t \leq 1$, a 1-parameter family of 2-forms. For each t, $w_t(0) = w(0)$ is non-degenerate, so that w_t is non-degenerate in some neighborhood of 0, for all $0 \leq t \leq 1$, which can be taken to be a ball. Now, $d(w_1 - w) = -dw = 0$, so by the Poincaré Lemma, $w_1 - w = d\alpha$, for some 1-form α, which we can choose such that $\alpha(0) = 0$. Define now a t-dependent vector field X_t by $i_{x_t} w_t = -\alpha$, which is possible since w_t is non-degenerate. Let χ_t be the time-dependent flow of χ_t with $\chi_0 = Id$. Then

$$\begin{aligned}\frac{d}{dt}(\tilde{\chi}_t w_t) &= \tilde{\chi}_t \frac{dw_t}{dt} + \tilde{\chi}_t (L_{x_t} w_t) \\ &= \tilde{\chi}_t (w_1 - w) + \tilde{\chi}_t (d i_{X_t} w_t + i_{X_t} dw_t) \\ &= \chi_t (w_1 - w - d\alpha) = 0.\end{aligned}$$

Putting $t = 1$, $\tilde{\chi}_1 w_1 = \tilde{\chi}_0 w_0 = w$. Thus, χ_1 gives a coordinate change transforming w to the constant form w_1. We can now use (B.24) to obtain (B.26). □

The coordinate system given by (B.26) defines a *symplectic* or *canonical* chart on (M, w). It is interesting to contrast this with a pseudo-Riemannian manifold (M, g) where the metric g can be made constant in a coordinate chart only when the curvature vanishes.

The fundamental result of interest to classical mechanics is the next theorem which shows that the cotangent bundle of every manifold carries a natural (or canonical) symplectic structure.

Theorem. *The cotangent bundle T^*M of every manifold M has a canonical symplectic structure.*

Proof. On the tangent bundle TT^*M of T^*M we have two projections, the ordinary tangent bundle projection $p_{T^*M} : TT^*M \to T^*M$ and the differential $(q_M^*) : TT^*M \to TM$ of $q_M : T^*M \to M$. If $\alpha \in TT^*M$, then $p_{T^*M}\alpha \in T^*M$ and $(q_{M^*})\alpha \in TM$. We can thus define canonically a 1-form T^*M by means of an element θ of T^*T^*M as follows

$$\langle \theta, \alpha \rangle = \langle (q_M)_*\alpha, p_{T^*M}\alpha \rangle. \tag{B.27}$$

That this defines a 1-form θ on T^*M is clear. We now consider its expression in local coordinates. Let $(q_i, p_i) = (q_1, \ldots, q_n,\ p_1, \ldots, p_n)$ be the standard local coordinates of T^*M. An element $\alpha \in TT^*M$ has local coordinate expression

$$\alpha = \sum a_i \frac{\partial}{\partial q_i} + \sum b_i \frac{\partial}{\partial p_i}.$$

In other words $\alpha = (q_i, p_i, a_i, b_i)$ and so $p_{T^*M}\alpha = (q_i, p_i)$. On the other hand, $(q_M)_*\alpha = (q, a_i)$. Therefore,

$$\langle p_{T^*M}\alpha, (q_M)_*\alpha \rangle = \sum p_i a_i = \langle \theta, \alpha \rangle$$

that is, $\theta = (q_i, p_i, p_i, 0)$. Its expression in local coordinates is thus

$$\theta = \sum p_i dq_i. \tag{B.28}$$

Now, let

$$w = -d\theta = \sum dq_i \wedge dp_i. \tag{B.29}$$

Then, w defines a non-degenerate, closed, 2-form and (T^*M, w) is, thus, a symplectic manifold. □

Thus, if M stands for the configuration space of a mechanical system, its phase space (which is nothing but the cotangent bundle T^*M) has a *canonical symplectic structure.* From now on, we will have this situation in the back of our mind, although many ideas and results remain valid for general symplectic manifolds.

Incidentally, the above result shows that the cotangent bundle of every manifold is orientable. That the tangent bundle of every manifold is also orientable is a matter of simple exercise.

Since w is non-degenerate on T^*M, we have an isomorphism between the vector fields $\chi(T^*M)$ and l-forms $\Lambda'(T^*M)$ on T^*M as follows. Let $\alpha \in \Lambda'(T^*M)$; define $X_\alpha \in \chi(T^*M)$ by

$$\alpha \mapsto X_\alpha, \quad i_{X_\alpha} w = \alpha.$$

Conversely, let $X \in \chi(T^*M)$, then define $\alpha_X \in \Lambda'(t^*M)$ by

$$\alpha \mapsto \alpha_x, \quad i_X w = \alpha_X.$$

Obviously, $\alpha_{X_\alpha} = \alpha$ and $X_{\alpha_X} = X$.

In local coordinates (q_i, p_i) of T^*M, we have

$$\alpha = \sum (a_i dq_i + b_i dp_i) \;\mapsto\; X_\alpha = \sum \left(b_i \frac{\partial}{\partial q_i} - a \frac{\partial}{\partial p_i} \right). \tag{B.30}$$

Definition. A vector field $X \in \chi(T^*M)$ is said to be locally Hamiltonian (system) if α_X is closed, and globally Hamiltonian if α_X is exact.

A characterizing property of locally Hamiltonian vector fields is the following proposition.

Proposition. *X locally Hamiltonian* $\Leftrightarrow L_X w = 0$.

Proof. $L_X w = i_X dw + d i_X w = d i_X w = d\alpha_X$. □

One can generate globally Hamiltonian vector fields from functions by taking the exterior derivative. That is, if $H : T^*M \to \mathbb{R}$, then X_{dH}, defined by $i_{X_{dH}} w = dH$, is globally Hamiltonian.

In local coordinates, let $H = H(q_i, p_i)$, then from (B.30)

$$X_{dH} = \sum \left(\frac{\partial H}{\partial p_i} \frac{\partial}{\partial q_i} - \frac{\partial H}{\partial q_i} \frac{\partial}{\partial p_i} \right). \tag{B.31}$$

The integral curves of X_{dH} are then precisely the Hamiltonian equations

$$\left. \begin{aligned} \dot{q}_i &= \frac{\partial H}{\partial p_i} \\ \dot{p}_i &= -\frac{\partial H}{\partial q_i}. \end{aligned} \right\} \tag{B.8}$$

Our geometric picture of the Hamiltonian formalism is then (T^*M, w, X_{dH}), i.e., the phase space T^*M, as the state space, with a symplectic structure w and a Hamiltonian vector field (or system) $X_{dH'}$ generated by a Hamiltonian function H.

The conservation of energy principle follows from the next proposition.

Proposition. *H is constant along the integral curves of X_{dH}.*

Proof.

$$\begin{aligned} L_{X_{dH}} H = X_{dH}(H) = dH(X_{dH}) &= (i_{X_{dH}} w)(X_{dH}) \\ &= w(X_{dH}, X_{dH}) = 0. \end{aligned}$$ □

Definition. The Poisson bracket (P.B.) of two functions $f, g : T^*M \to \mathbb{R}$ is given by

$$\{f, g\} = w(X_{df}, X_{dg}). \tag{B.32}$$

In other words, $\{f, g\} = -i_{X_{df}} i_{X_{dg}} w = i_{X_{dg}} i_{X_{df}} w$. We have another P.B. characterization. Since

$$-L_{X_{df}} g = -i_{X_{df}} dg - d i_{X_{df}} g = -i_{X_{df}} dg = -i_{X_{df}} i_{X_{dg}} w,$$

we have

$$\{f, g\} = -L_{X_{df}} g = L_{X_{dg}} f = X_{dg} f = -X_{df} g. \tag{B.33}$$

In local coordinates,

$$-X_{df} = -\sum \left(\frac{\partial f}{\partial p_i} \frac{\partial}{\partial q_i} - \frac{\partial f}{\partial q_i} \frac{\partial}{\partial p_i} \right),$$

so

$$\{f, g\} = \sum \left(\frac{\partial f}{\partial q_i} \frac{\partial g}{\partial p_i} - \frac{\partial g}{\partial p_i} \frac{\partial f}{\partial q_i} \right). \tag{B.34}$$

Proposition. *A function $f : T^*M \to \mathbb{R}$ is a first integral of the Hamiltonian system X_{dH}, i.e., f is constant along the integral curves of X_{dH}, iff $\{f, H\} = 0$.*

Proof. From (B.33), $\{f, H\} = L_{X_{dH}} f$. So, $\{f, H\} = 0 \Leftrightarrow L_{X_{dH}} f = 0$. Thus, the energy conservation principle also follows from the definitions of P.B., because $\{H, H\} = 0$. □

Let χ_t be the flow of the Hamiltonian system X_{dH}. Then, for any function f,

$$\begin{aligned}\frac{d}{dt}(f \circ \chi_t) &= \frac{d}{dt}(\tilde{\chi}_t f) = \tilde{\chi}_t L_{X_{dH}} f \\ &= L_{\tilde{\chi}_t X_{dH}}(\tilde{\chi}_t f) = L_{X_{dH}}(f \circ \chi_t) = \{f \circ \chi_t, H\}. \quad \text{(B.35)}\end{aligned}$$

In local coordinates, Eq. (B.35) is equivalent to Eq. (B.9), that is, equations of motion in P.B. form.

The following properties of the P.B. are easily checked:

(i) $\{f, g+h\} = \{f, g\} + \{f, h\}$,
(ii) $\{f, \lambda g\} = \lambda\{f, g\}$, $\lambda \in \mathbb{R}$,
(iii) $\{f, g\} = -\{g, f\}$,
(iv) $\{f, \{g, h\}\} + \{g, \{h, f\}\} + \{h, \{f, g\}\} = 0$ (Jacobi identity),
(v) $\{f, gh\} = h\{f, g\} + g\{f, h\}$.

They show that the (vector) space of functions $F(t^*M)$ on T^*M, together with the P.B., form a Lie algebra. The Jacobi identity (iv) can also be rewritten as

$$X_{d\{f,g\}} = -[X_{df}, X_{dg}] \quad \text{(B.36)}$$

where the right-hand side is the usual Lie bracket of the vector fields X_{df}, X_{dg}.

Problem. Given a proof of (B.36) and, thus, of the Jacobi identity (iv).

(B.36) now implies the following:

Proposition. *The mapping* $F(T^*M) \to \chi(T^*M)$, *given by* $f \mapsto X_{df}$, *is a Lie algebra homomorphism.*

Symplectic Diffeomorphisms or Canonical Transformations.

Definition. Let (N, w) and (N', w') be two general symplectic manifolds and $\phi : N \to N'$ a diffeomorphism. Then, ϕ is said to be a symplectic diffeomorphism or a canonical transformation if $\phi^* w' = w$.

Although we will be concerned mainly with the case $N = N' = T^*M$, $w = w'$, $\phi : T^*M \to T^*M$, the next two propositions are formulated in terms of general symplectic manifolds. They give two different characterizations of canonical transformations.

Proposition. *Let $\phi : N \to N'$ be a diffeomorphism of (N, w) onto (N', w'). Then, ϕ is a canonical transformation iff*

(i) $\phi^* X_{dH} = X_{d(\phi^* H)}$ *for all* $H \in F(N')$,
(ii) $\{\phi^* f,\ \phi^* g\} = \phi^*\{f, g\}$ *for all* $f, g \in F(N')$.

Proof.

(i)

$$\begin{aligned} i_{X_{dH}} w' = dH \Rightarrow\ & \phi^*(i_{X_{dH}} w') = \phi^*(dH) \\ \text{or}\quad & i_{\phi^* X_{dH}}(\phi^* w') = d(\phi^* H) \\ \text{or}\quad & i_{\phi^* X_{dH}}(\phi^* w') = i_{X_{d(\phi^* H)}} w. \end{aligned}$$

So, if $\phi^* w' = w$, then $\phi^* X_{dH} = X_{d(\phi^* H)}$. Conversely, suppose $\phi^* X_{dH} = X_{d(\phi^* H)}$. Then, $i_{X_{d(\phi^* H)}}(\phi^* w') = i_{X_{d(\phi^* H)}} w$. Or $\phi^* w' = w$, since every vector at $p \in N$ is of the form $X_{d(\phi^* H)}(p)$ for some H on N'.

(ii) We have $\phi^*\{f, g\} = \phi^* L_{X_{dg}} f = L_{\phi^* X_{dg}} \phi^* f$, so

$$\begin{aligned} \phi^*\{f, g\} = \{\phi^* f,\ \phi^* g\} &\Leftrightarrow \phi^* X_{dg} = X_{d(\phi^* g)} \\ &\Leftrightarrow \phi^* w' = w \text{ from (i).} \end{aligned}$$

□

Thus the Hamiltonian equations and the P.B.'s are preserved under a canonical transformation $\phi : T^*M \to T^*M$.

It is now easy to see that if χ_t is the flow of X_{dH}, then χ_t is a (local) canonical transformation for any t. This follows from the fact that X_{dH} is a Hamiltonian vector field and thus $L_{X_{dH}} w = 0$, and hence, $\tilde{\chi}_t w = w$. Not only χ_t preserves w, it also preserves $w^2, w^3, \ldots, w^n$. Now, w^n is a volume element of T^*M and in local coordinates (q_i, p_i),

$$w^n = \pm n!\, dq_1 \wedge \cdots \wedge dq_n \wedge dp_1 \wedge \cdots \wedge dp_n. \tag{B.37}$$

We have thus proved the so-called Liouville Theorem.

Liouville Theorem. $\chi_{t'}$, *the flow of* X_{dH}, *preserves* w^n, *the volume element of the phase space, i.e.,* $L_{X_{dH}} w^n = 0$.

In general, $\alpha \in \Lambda(T^*M)$ is called an *invariant form* of X_{dH} iff $L_{X_{dH}}\alpha = 0$, i.e., α is constant along the flow lines of X_{dH}. When $\alpha = f$, a function, it is called a *first integral* of the Hamiltonian system. For example, $w, w^2, \ldots, w^n$ are invariant forms and H a first integral of X_{dH}. Invariant forms have the following integral property.

Poincaré's Theorem. *Suppose* X_{dH} *is a complete vector field,* χ_t *its flow,* V *a compact oriented* 2*-manifold with boundary and* $\phi : V \to T^*M$ *a smooth map. Then*

$$\int_V (\chi_t \circ \phi)^* w = \int_V \phi^* w \tag{B.38}$$

i.e., independent of t.

Proof. Since X_{dH} is complete, χ_t is a global flow on T^*M and since V is compact and orientable, the integrals are well defined. And since w is an invariant form, $(\chi_t \circ \phi)^* w = (\phi^* \circ \chi_t^*) w = \phi^* w$. □

The canonical transformations obviously form a group. We now consider an important *subgroup* of the group of all canonical transformations, namely, those which preserve the canonical *l*-form θ on

T^*M, i.e., those diffeomorphisms $\psi : T^*M \to T^*M$, such that $\tilde{\psi}\theta = \theta$. Such canonical transformations are called *homogeneous canonical transformations*. The next theorem shows how such homogeneous transformations are generated from diffeomorphisms of the base manifolds M.

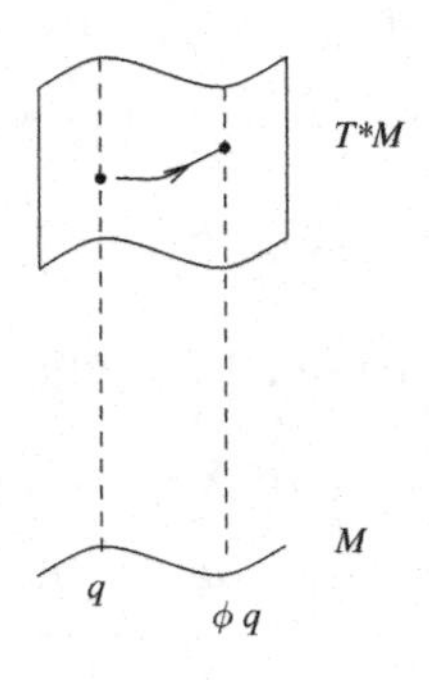

Theorem. *Let* $\phi : M \to M$ *be a diffeomorphism. Define a map* $\psi : T^*M \to T^*M$ *by* $(q, p) \mapsto (\phi q, (\phi^{-1})^* p)$. *Here* $q \in M$, $p \in T_q^*(M)$, $(\phi^{-1})^*_{\phi q} : T_q^*(M) \to T^*_{\phi q}(M)$ *is the pull-back from* q *to* ϕq. *Then* ψ *is a diffeomorphism of* T^*M *such that*

(i) $q_M \circ \psi = \phi \circ q_M$ *and*
(ii) $\tilde{\psi}\theta = \theta$, *where* θ *is the canonical l-form on* T^*M.

Proof. (i) follows from the definition of ψ. We now check (ii). Let $X_{(q,p)} \in T_{(q,p)}(T^*M)$, i.e., a tangent vector at the point (q, p) of T^*M. Note that $\theta_{(q,p)} \in T^*_{(q,p)}(T^*M)$. So consider

$$\langle X_{(q,p)}, (\tilde{\psi}\theta)_{(q,p)} \rangle = \langle X_{(q,p)}, \psi^*_{(q,p)} \theta_{\psi(q,p)} \rangle,$$

by the definition of $\tilde{\psi}$,

$$= \langle \psi_{*(q,p)} X_{(q,p)},\ \theta_{\psi(q,p)} \rangle$$
$$= \langle (q_M)_{*\psi(q,p)} \psi^*_{(q,p)} X_{(q,p)},\ (\phi^{-1})^*_{\phi q} p \rangle,$$

by the definition of θ,

$$= \langle \phi_{*q} (q_M)_{*(q,p)} X_{(q,p)},\ (\phi^{-1})^*_{\phi q} p \rangle \quad \text{by (i)}$$
$$= \langle (q_M)_{*(q,p)} X_{(q,p)}, p \rangle, \quad \text{since } (\phi^{-1})^*_{\phi q} \cdot \phi_{*q} = Id.$$
$$= \langle X_{q,p)}, \theta_{(q,p)} \rangle, \quad \text{by the definition of } \theta \text{ again.}$$

□

Thus, every diffeomorphism of the configuration space M can be lifted (canonically) to a homogeneous canonical transformation of the phase space. Such homogeneous canonical transformations are therefore called *point transformations.*

In local coordinates, if $q_i \to \bar{q}_i(q)$ defines $\phi : M \to M$ locally, then $\psi : T^*M \to T^*M$ is given by $(q_i, p_i) \to (\bar{q}_i, \bar{p}_i)$ where $p_i = \sum (\frac{\partial \bar{q}_j}{\partial q_i})^{\bar{p}_j}$.

The above theorem remains valid if $\phi_t : M \to M$ is now a local l-parameter group of local diffeomorphisms of M, thereby giving a local l-parameter group of local diffeomorphisms ψ_t of T^*M. Hence we have the following *canonical lift theorem* for vector fields on M.

Theorem. *Let Y be a vector field on M. Then, there is a unique vector field Z on T^*M, such that*

(i) $(q_M)_* \circ Z = Y \circ q_M$, *i.e., the diagram*

$$\begin{array}{ccc} T^*M & \xrightarrow{Z} & T\,T^*M \\ \Big\downarrow q_M & & \Big\downarrow (q_M)_* \\ M & \xrightarrow{X} & TM \end{array} \quad \textit{commutes}$$

(ii) $L_Z \theta = 0$.

Proof. Let ϕ_t be the flow of Y and let ψ_t, its canonical lift on T^*M, generate Z. Then, from the previous theorem, (i) and (ii) are satisfied. To see uniqueness, we consider their expressions in local coordinates (q_i) and (q_i, p_i). Let

$$Y = \sum a_i \frac{\partial}{\partial q_i} \quad \text{and} \quad Z = \sum \left(b_i \frac{\partial}{\partial q_i} + c_i \frac{\partial}{\partial p_i} \right).$$

Then, the condition (i) $\Rightarrow$ $b_i = a_i$.

So, $Z = (a_i \frac{\partial}{\partial q_i} + c_i \frac{\partial}{\partial p_i})$.

Now, $\theta = \sum p_i dq_i$. Therefore, condition (ii) implies

$$\begin{aligned} 0 = L_Z\theta &= i_Z d\theta + d i_Z \theta \\ &= \sum c_i dq_i - \sum b_i dp_i + \sum b_i dp_i + \sum p_i da_i, \\ \text{or} \quad c_i &= -\sum p_j \frac{\partial a_j}{\partial q_i}. \end{aligned}$$

So, the unique canonical lift of $Y = \sum a_i \frac{\partial}{\partial q_i}$ is

$$Z = \sum_i a_i \left[\frac{\partial}{\partial q_i} - \left(\sum_j p_j \frac{\partial a_j}{\partial q_i} \right) \frac{\partial}{\partial p_i} \right]. \tag{B.39}$$

□

For example, if $Y = \frac{\partial}{\partial q_i}$, then $Z = \frac{\partial}{\partial q_i}$.

Condition (ii) shows that Z is a locally Hamiltonian vector field on T^*M. It is, in fact, a *globally* Hamiltonian vector field, because

$$\begin{aligned} L_Z\theta = 0 &\Rightarrow -i_Z d\theta = d i_Z \theta \\ \text{or} \quad i_Z w &= d\theta(Z). \end{aligned} \tag{B.40}$$

In other words, Y gives rise to a function $\theta(Z)$ on T^*M, which in turn, generates the globally Hamiltonian vector field Z.

For example, if $Y = \frac{\partial}{\partial q_i}$ on M, then $\theta(Z) = p_i$ on T^*N.

We shall now apply these results to derive an important relationship between *symmetries* of a Hamiltonian system and *conservation laws.* Let G be a (Lie) transformation group acting on M, that is, there is a smooth map $MXG \to M$, given by $(m, g) \mapsto mg$, such that (i) $\forall g \in G, m \mapsto mg$ is a diffeomorphism of M into M, and (ii) $m(g_1 g_2) = (mg_1)g_2$, for all $m \in M$, $g_1, g_2 \in G$. The action of each l-parameter subgroup of G gives rise to a l-parameter group of diffeomorphisms of M and hence, a vector field on M. More precisely, one has a Lie algebra homomorphism of the Lie algebra of G into the Lie algebra $\Xi(M)$ of vector fields on M. Thus, a k-dimensional Lie group G acting on M gives rise to k vector field y_α $(\alpha = 1, \ldots, k)$ on M. These are the so-called *infinitesimal generators* of G. For example, the translation group acting on $\mathbb{R}^n$ has the generators $Y_i = \frac{\partial}{\partial q_i}$ and the rotation group, $Y_{ik} = q_i \frac{\partial}{\partial q_k} - q_k \frac{\partial}{\partial q_i}$.

Definition. G is said to be a symmetry group of the Hamiltonian system (T^*M, w, X_{dH}) if the canonical lift Z of each of its infinitesimal generators Y leaves the Hamilton H invariant, i.e., H is invariant under the flow of Z of $L_Z H = Z(H) = 0$.

Proposition. *If G is a symmetry group of the Hamiltonian system, then $\theta(Z)$ is a constant of motion, i.e., a first integral.*

Proof. We have

$$\begin{aligned} Z(H) &= dH(Z) = (i_{X_{dH}} w)(Z) = w(X_{dH}, Z) \\ &= -d\theta(X_{dH}, Z) = -\{X_{dH}\theta(Z) - Z\theta(X_{dH}) - \theta([X_{dH}, Z])\} \\ &= -\{X_{dH}\theta(Z) - (L_Z\theta)(X_{dH})\} = -X_{dH}\theta(Z). \end{aligned}$$

Thus, $Z(H) = 0 \Rightarrow X_{dH}\theta(Z) = 0$. □

For example, if the Hamiltonian is invariant under translations, we have,

$$Y_i = \frac{\partial}{\partial q_i}$$
$$Z_i = \frac{\partial}{\partial q_i}, \quad \theta(Z_i) = p_i$$

so that the momentum is conserved. Similarly, if the Hamiltonian is invariant under the rotation group, the angular momentum is conserved.

Lagrangian Systems. We now turn to the Lagrangian formalism whose state space is the tangent bundle TM, i.e., position-velocity space.

One should keep in mind that both the tangent and cotangent bundles, as their names indicate, besides being manifolds, have the structure of vector bundles. A collection (E, p, B, F), where E, B are manifolds, F a vector space and $p : E \to B$ a subjective map, is called a *vector bundle* if $\forall\, b \in B$, there exists a neighborhood U of b and a map $\phi : p^{-1}(U) \to UXF$, such that (i) ϕ is a diffeomorphism (ii) $p_1 \circ \phi = p$, where p_1 is the canonical projection $UXF \to U$ in the first variable, and (iii) $\phi^{-1}(b, f + f') = \phi^{-1}(b, f')$, $\phi^{-1}(b, \lambda f) = \lambda \phi^{-1}(b, f)$, $\forall\, f, f' \in F$, $\lambda \in \mathbb{R}$. (Note that $\phi^{-1}(\{b\}XF) \simeq F$.)

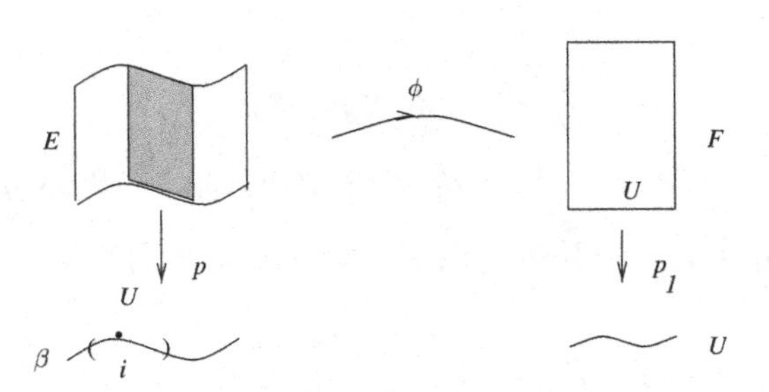

For each $b \in B$, $p^{-1}(b)$ is thus diffeomorphic to F, the *fiber*. E is the *total* space or *bundle* and B, the *base manifold*. Thus, $(TM, p_M, M, \mathbb{R}^n)$ and $(T^*M, q_M, M, \mathbb{R}^n)$ are vector bundles over the same base manifold M and fiber $\mathbb{R}^n$.

Now, a map $f : E \to E'$, where (E, p, B, F), (E', p', B, F) are vector bundles over the same manifold B and fiber F is said to be a vector bundle map if (i) $p' \circ f = p$ and (ii) f restricted to each fiber is a *linear* map.

$$\begin{array}{ccc} E & \xrightarrow{f} & E' \\ p \searrow & & \swarrow p' \\ & B & \end{array}$$

Now let $L : TM \to \mathbb{R}$ be a smooth function. Our objective is to derive a map $TM \to T^*M$, which is fiber preserving but not necessarily a vector bundle map.

$$\begin{array}{ccc} TM & \xrightarrow{FL} & T^*M \\ \downarrow L & & \\ \mathbb{R} & & \end{array}$$

Let $m \in M$, $X_m \in T_m(M)$ and L_m, the restriction of L to $p_M^{-1}(m)$, the fiber over m, so that $L_m : T_m(M) \to R$ and thus, the derivative of L_m is given by $DL_m : T_m(M) \to L(T_mM, \mathbb{R})$, the space of linear maps from $T_mM \to \mathbb{R}$. Hence, $DL_m(X_m) \in T_m^*(M)$.

Definition. The fiber derivative FL of a map $L : TM \to \mathbb{R}$ is the map $FL : TM \to T^*M$ given by $X_m \mapsto DL_m(X_m)$.

FL is a fiber preserving map, but not a vector bundle map because it is not necessarily linear on the fibers. In local coordinates (q_i, v_i)

of TM and (q_i, p_i) of T^*M, if $L = L(q_i, v_i)$, then

$$FL : (q_i, v_i) \mapsto \left(q_i, p_i = \frac{\partial L}{\partial v_i}\right). \tag{B.41}$$

With the help of FL, we can now pull back the canonical symplectic 2-form w on T^*M into TM. Thus,

$$w_L = (FL)^* w \tag{B.42}$$

is a 2-form on TM. However, in general, w_L would not be symplectic. If L is such that FL is a *local diffeomorphism*, then L is said to be a *regular Lagrangian* and in that case, w_L is a symplectic 2-form on TM. (Note that w_L is always *closed*, because $dw_L = d\{(FL)^* w\} = (FL)^* dw = 0$. The regularity condition is required to make w_L non-degenerate.) Assuming then that L is regular, (TM, w_L) *becomes a symplectic manifold.*

In local coordinates, from (B.41) we have

$$\begin{aligned} w_L &= (FL)^* w = (FL)^* \sum dq_i \wedge dp_i = \sum (FL)^* dq_i \wedge (FL)^* dp_i \\ &= \sum d((FL)^* q_i) \wedge d((FL)^* p_i) = \sum dq_i \wedge \left(\frac{\partial L}{\partial v_i}\right) \end{aligned}$$

$$\text{or} \quad w_L = \sum \frac{\partial^2 L}{\partial q_j \partial v_i} dq_i \wedge dq_j + \sum \frac{\partial^2 L}{\partial v_j \partial v_i} dq_i \wedge dv_j. \tag{B.43}$$

Note the regularity of $L \Leftrightarrow w_L^n \neq 0 \Leftrightarrow \det(\frac{\partial^2 L}{\partial v_j \partial v_i}) \neq 0$. One can also write

$$w_L = -d\theta_L, \quad \text{where} \quad \theta_L = (FL)^* \theta \tag{B.44}$$

and in local coordinates

$$\theta_L = \sum \frac{\partial L}{\partial v_i} dq_i. \tag{B.45}$$

We now have to find the dynamical vector field for the Lagrangian formalism. Note that since $FL : TM \to T^*M$, if $X_m \in T_m(M)$, then $FL(X_m)X_m \in \mathbb{R}$.

Definition. The action $A : TM \to \mathbb{R}$ is given by $X_m \to FL(X_m) \cdot X_m$ and the energy function is $E = A - L$.

In local coordinates, $A = \sum v_i \frac{\partial L}{\partial v_i}$. In other words, $A = V(L)$, where V is given by (B.22). Hence, $E = \sum v_i \frac{\partial L}{\partial v_i} - L$, which is nothing but the energy (or Hamiltonian) expressed as a function of (q_i, v_i).

Definition. The dynamical Lagrangian vector field X_{dE} is given by

$$i_{X_{dE}} w_L = dE. \tag{B.46}$$

And, by a Lagrangian formalism, we meant the triple (TM, w_L, X_{dE}).

Theorem. *X_{dE} is a second-order O.D.E. on M.*

Proof. In local coordinates, let

$$X_{dE} = \sum a_i \frac{\partial}{\partial q_i} + \sum b_i \frac{\partial}{\partial v_i}.$$

Then from (B.43),

$$\begin{aligned} i_{X_{dE}} w_L = & \frac{\partial^2 L}{\partial v_j \partial v_i} a_i dv_j - \sum \frac{\partial^2 L}{\partial v_j \partial v_i} b_i dq_j \\ & + \sum \frac{\partial^2 L}{\partial q_i \partial v_j} a_j dq_i - \sum \frac{\partial^2 L}{\partial q_j \partial v_i} a_j dq_i. \end{aligned}$$

On the other hand,

$$dE = \sum \frac{\partial^2 L}{\partial v_j \partial v_i} v_i dv_j + \sum \frac{\partial^2 L}{\partial q_j \partial v_i} v_i dq_j - \sum \frac{\partial L}{\partial q_j} dq_j.$$

Hence,

$$i_{X_{dE}} w_L = dE \Leftrightarrow \begin{cases} \sum_i \frac{\partial^2 L}{\partial v_j \partial v_i} a_i = \sum_i \frac{\partial^2 L}{\partial v_j \partial v_i} v_i, & \text{(B.47)} \\ -\sum_i \frac{\partial^2 L}{\partial v_j \partial v_i} b_i = \sum_i \frac{\partial^2 L}{\partial v_j \partial v_i} v_i - \frac{\partial L}{\partial q_j}. & \text{(B.48)} \end{cases}$$

Note that since L is regular, (B.47) implies that $a_i = v_i$. □

Corollary. *The integral curves of X_{dE} satisfy the Lagrange equations* (B.4).

Proof. The integral curves of X_{dE} are given by

$$\left. \begin{aligned} \frac{dq_i}{dt} &= a_i = v_i \\ \frac{dv_i}{dt} &= b_i. \end{aligned} \right\} \quad \text{(B.49)}$$

Now, (B.48) and (B.49) $\Rightarrow \sum_i \frac{\partial^2 L}{\partial q_i \partial v_j} \frac{dq_i}{dt} + \sum_i \frac{\partial^2 L}{\partial v_j \partial v_i} \frac{dv_i}{dt} - \frac{\partial L}{\partial q_j} = 0$ or

$$\frac{d}{dt}\left(\frac{\partial L}{\partial v_j}\right) - \frac{\partial L}{\partial q_j} \equiv \frac{d}{dt}\left(\frac{\partial L}{\partial \dot{q}_j}\right) - \frac{\partial L}{\partial q_j} = 0.$$ □

Remark. Because of the way it is defined, a second-order O.D.E. is possible only on TM and not on T^*M. This explains why the Lagrangian equations are second-order O.D.E., while the Hamiltonians are of first-order.

A special case of Lagrangians are the homogeneous ones, which are important in mechanics. L is said to be *homogeneous of degree* k if $L_V L \equiv V(L) = kL$, where $V = \sum v_i \frac{\partial}{\partial v_i}$, the Liouville vector field (B.22) on TM. In other words, $V(L) = \sum v_i \frac{\partial L}{\partial v_i} = kL$, that is, L is homogeneous of degree k in the v_i's. In general, a form α is homogeneous of degree k if $L_V \alpha = k\alpha$.

Proposition. *If L is homogeneous of degree k, then X_{dE} is a spray.*

Proof. We have already seen that X_{dE} is a second-order O.D.E. on M. We only have to show that $[V, X_{dE}] = X_{dE}$. This can be seen by establishing that $i_{[V,X_{dE}]}w_L = i_{X_{dE}}w_L$ if $V(L) = kL$.

$$E = A - L = V(L) - L = (k - l)L$$

so

$$i_{X_{dE}}w_L = dE = (k - l)dL.$$

Now

$$\begin{aligned} i_{[V,X_{dE}]}w_L &= (L_V i_{X_{dE}} - i_{X_{dE}}L_V)w_L \\ &= L_V i_{X_{dE}}w_L - i_{X_{dE}}L_v w_L, \end{aligned}$$

and

$$L_V i_{X_{dE}}w_L = L_V dE = L_V(k - l)dL = (k - l)dL_V L = k(k - l)dL.$$

Since $\theta_L = \sum \frac{\partial L}{\partial v_i}dq_i$ and L is homogeneous of degree k in V_i, θ_L is homogeneous of degree $k - 1$, i.e., $L_V\theta_L = (k - l)\theta_L$ so $L_V w_L = -L_V d\theta_L = -dL_V\theta_L = -d(k - l)\theta_L = (k - l)w_L$. Therefore,

$$\begin{aligned} i_{X_{dE}}L_V w_L &= i_{X_{dE}}(k - l)w_L = (k - l)i_{X_{dE}}w_L \\ &= (k - l)dE = (k - l)(k - 1)dL \end{aligned}$$

and so

$$i_{[V,X_{dE}]}w_L = k(k - l)dL - (k - l)(k - l)dL = (k - l)dL = i_{X_{dE}}w_L.$$

□

Legendre Transformation and Passage to the Hamiltonian Formalism. One can pass from a Lagrangian formalism (TM, w_L, X_{dE}) to a corresponding Hamiltonian formalism

(T^*M, w, X_{dH}) by means of the so-called Legendre transformation. Although $T_m(M)$ and $T_m^*(M)$ are isomorphic to each other, both being n-dimensional vector spaces, there is no *canonical* isomorphism between the two spaces. The Lagrangian L through the fiber derivative FL provides such an isomorphism in the following case.

Definition. L is said to be hyper-regular if $FLa : TM \to T^*M$ is a (global) diffeomorphism.

Since $w_L = (FL)^*w$, FL is a symplectic diffeomorphism if L is hyper-regular and thus preserves Hamiltonians and P.B.'s. The map $FL : TM \to T^*M$ *is called a Legendre transformation.*

Theorem. *Let* $L : TM \to \mathbb{R}$ *be hyper-regular, and* E *the energy function of* L*. Let* $E \circ (FL)^{-1} = H : T^*M \to \mathbb{R}$*; then* X_{dE} *given by* $i_{X_{dE}} w_L = dE$ *and* X_{dH} *given by* $i_{X_{dH}} w = dH$ *are related by* $(FL)_* X_{dE} = X_{dH}$*, so that the integral curves of* X_{dE} *on* TM *are mapped by* FL *onto the integral curves of* X_{dH} *on* T^*M.

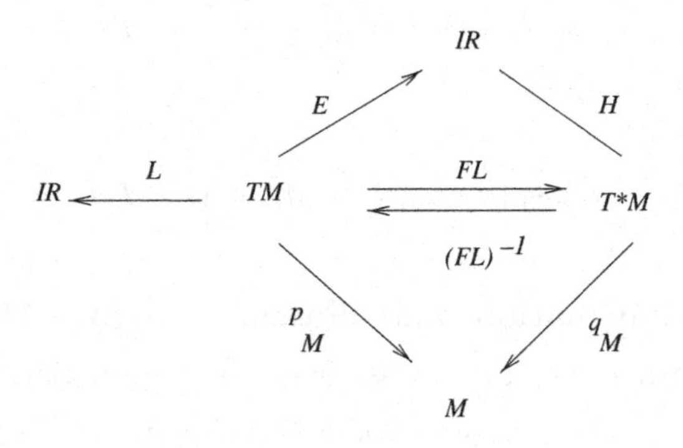

Proof. Since $FL : TM \to T^*M$ is a diffeomorphism, not only $w_L = (Fl)^*w$, but also $w = (FL)_*w_L$. Also since $H = E \circ (FL)^{-1}$, we have $dH = (FL)_*dE$. Hence,

$$\begin{aligned} i_{(FL)_*X_{dE}}w = i_{(FL)_*X_{dE}}(FL)_*w_L = (FL)_*(i_{X_{dE}}w_L) = (FL)_*dE \\ = dH = i_{X_{dH}}w \end{aligned}$$

so that $(FL)_*X_{dE} = X_{dH}$. □

Incidentally, since $p_M = q_M \circ FL$, the base integral curves on the configuration manifold M are identical and thus the two formalisms, Lagrangian and Hamiltonian, become identical.

Contact Manifolds ([7]) and Hamilton–Jacobi Theory. For time-dependent (i.e., non-autonomous) Hamiltonians, it is necessary to consider the *extended phase space* $T^*M \times \mathbb{R}$, where the second variable plays the role of time. Now $T^*M \times \mathbb{R}$, being an odd-dimensional manifold, cannot have a symplectic structure. The appropriate structure for the extended phase space is *contact* structure, which we consider very briefly.

Definition. A contact manifold $(N, \hat{w})$ consists of an odd-dimensional manifold N, together with a closed 2-form $\hat{w}$ of maximal rank.

An exact contact manifold $(N, \hat{\theta})$ consists of a $(2n+1)$-dimensional manifold N together with a 1-form $\hat{\theta}$ such that

$$\hat{\theta} \wedge (d\hat{\theta})^n \neq 0 \quad \text{(i.e., a volume element on } N). \tag{B.50}$$

Thus, if $(N, \hat{\theta})$ is an exact contact manifold, then $(N, d\hat{\theta})$ is a contact manifold and a contact manifold is locally an exact contact manifold. Darboux's Theorem can be generalized to exact contact manifolds to give existence of local coordinates $(x_1, \ldots, x_{2n+1})$ on $U \subset N$, such that

$$\left.\begin{aligned} \hat{\theta}_{|_U} &= dx_1 + x_2 dx_3 + \cdots + x_{2n} dx_{2n+1} \\ \text{and} \quad \hat{\theta} \wedge (d\hat{\theta})^n|_U &= \pm n! dx_1 \wedge \cdots \wedge dx_{2n+1}. \end{aligned}\right\} \tag{B.51}$$

On $(N, \hat{\theta})$, there exists a characteristic vector field Y such that

$$\hat{\theta}(Y) = 1 \quad \text{and} \quad i_Y(d\hat{\theta}) = 0. \tag{B.52}$$

This can be seen by taking $Y = \frac{\partial}{\partial x_1}$ in the above coordinates. Note also that $L_Y(\hat{\theta}) = L_Y(d\hat{\theta}) = 0$.

For an example of exact contact manifolds from mechanics, consider the Hamiltonian formalism $(T^*M,\ w = -d\theta, X_{dH})$, where $H : T^*M \to \mathbb{R}$. Let $h \in \mathbb{R}$ and suppose $H^{-1}(h) \neq \phi$. The connected set of regular points of $H^{-1}(h)$ (i.e., where $dH \neq 0$) constitute a *regular energy surface* of H and is a submanifold of TM of dimension $2n - 1$. Since H is constant along the integral curves of $X_{dH'}$, every integral curve of X_{dH} that meets a regular energy surface $H^{-1}(h)$ lies entirely within it. Now, let $i : H^{-1}(h) \to T^*M$ be the inclusion map. Then, $(H^{-1}(h),\ \hat{\theta} = i^*\theta)$ is an exact contact manifold.

Next, consider the extended phase space $T^*M \times \mathbb{R}$ and let $\rho : T^*M \times \mathbb{R} \to T^*M$ be the canonical projection of the first variable, $t : T^*M \times \mathbb{R} \to \mathbb{R}$ the canonical projection of the second variable, $H : T^*M \times \mathbb{R} \to \mathbb{R}$, a time-dependent Hamiltonian, $\hat{w} = \rho^* w$, where w is the symplectic 2-form on TM. Now set

$$w_H = \hat{w} + dH \wedge dt = \rho^* w + dH \wedge dt. \tag{B.53}$$

Then, $dw_H = 0$ and w_H is of maximal rank and so $(T^*M \times \mathbb{R}, w_H)$ is a contact manifold, and $(T^*M \times \mathbb{R}, \theta_H)$, where $\theta_H = \rho^*\theta - Hdt$, is an exact contact manifold if $H \neq 0$ on $T^*M \times \mathbb{R}$. We have $w_H = -d\theta_H$.

By a time-dependent canonical (or *contact*) transformation, we mean a diffeomorphism $\phi : T^*M \times \mathbb{R} \to T^*M \times \mathbb{R}$ such that (i) $\tilde{\phi} t = t$, i.e., ϕ preserves time, and (ii) $\theta_H - \tilde{\phi}\theta_H$ is *closed*. Note that (ii) implies $w_H = \tilde{\phi} w_H$, that is, ϕ preserves the contact structure of $(T^*M \times \mathbb{R}, w_H)$.

A function $F : T^*M \times \mathbb{R} \to \mathbb{R}$ is called a *generating* function for ϕ if $\theta_H - \tilde{\phi}\theta_H = dF$. If ϕ is a contact transformation, then *locally* such generating functions exist. In that case (since $\tilde{\phi} dt = dt$),

$$\rho^*\theta - \phi\rho^*\theta - (H - \tilde{\phi} H)dt = dF \tag{B.54}$$

which is equivalent to (B.14) in local coordinates. Hence

$$-(H - \tilde{\phi} H) = \frac{\partial F}{\partial t}.$$

If ϕ can be chosen such that $\tilde{\phi} H = \text{const.}$, we say ϕ transforms H to *equilibrium*. Then,

$$H + \frac{\partial F}{\partial t} = \text{const.} \tag{B.55}$$

which is the Hamilton–Jacobi equation.

References

[1] E. T. Whittaker, *A Treatise on the Analytical Dynamics of Particles and Rigid Bodies*, 4th edn. (Cambridge University Press, 1959).

[2] H. Goldstein, *Classical Mechanics* (Addison Wesley, Mass., 1950).

[3] V. I. Arnold, *Mathematical Methods of Classical Mechanics*, Springer Graduate Texts in Mathematics, No. 60 (Springer Verlag, New York, 1978).

[4] G. W. Mackey, *Mathematical Foundations of Quantum Mechanics* (Benjamin-Cummings, Mass., 1963).

[5] S. Kobayashi and K. Nomizu, *Foundations of Differential Geometry*, Vol. 1 (Wiley, New York, 1963).

[6] C. Godbillon, *Géometrie differentielle et mécanique analytique* (Hermann, Paris, 1969).

[7] R. Abraham and J. E. Marsden, *Foundations of Mechanics*, 2nd edn. (Benjamin-Cummings, Mass., 1978).

[8] A. Weinstein, *Lectures on Symplectic Manifolds*, C.B.M.S. Conference Series, No. 29 (Amer. Math. Soc., 1977).

A fairly exhaustive bibliography is given in [7] which is the definitive reference on this subject. Some other illustrative references are given below.

[9] V. I. Arnold and A. Avez, *Théorie ergodique des systemes dynamique* (Gauthiers-Villars, Paris, 1967).

[10] H. Flanders, *Differential Forms* (Academic Press, New York, 1963).

[11] J. Klein, *Ann. Inst. Fourier* **12** (1962) 1–124.

[12] J. E. Marsden, *Applications of Global Analysis in Mathematical Physics* (Publish or Perish, Boston, 1974).

[13] G. Reeb, *Acad. Roy. Belg. Cl. Sci. Mém. Coll.* 8° **27** (1952), 64 pp.

[14] J. M. Souriau, *Structures des systèmes dynamiques* (Dunod, Paris, 1970).

[15] S. Sternberg, *Lectures on Differential Geometry* (Prentice Hall, N.J., 1963).

[16] W. Thirring, *Classical Dynamical Systems — A Course in Mathematical Physics I* (Springer-Verlag, New York, 1978).

APPENDIX C
MAPLE PROGRAM 1

The following program in MAPLE checks that the 4-metric in Sec. 1.8 is really a solution of the vacuum Einstein field equations.

```
with(tensor):
coords:=[x,y,z,t]:
# Define the covariant metric tensor:
g:=array(symmetric,sparse,1 .. 4, 1 .. 4):
a:=1:b:=0:c:=0:d:=1:
g[1,1] := 2*exp(-r*t)*(a*cos(r*3^(1/2)*t)- b*sin(r*3^(1/2)*t)):
g[2,2]:=2*exp(-r*t)*(c*cos(r*3^(1/2)*t)-d*sin(r*3^(1/2)*t)):
A:=((c^2+d^2)^(1/2)* (a^2+b^2)^(1/2)+ a*c-b*d)^(1/2)/ sqrt(2):
B:=((c^2+d^2)^(1/2)* (a^2+b^2)^(1/2)
+ b*d-a*c)^(1/2)*signum(a*d+b*c)/ sqrt (2):
g[1,2]:=2*exp(-r*t)*(A*cos(r*3^(1/2)*t)-B*sin(r*3^(1/2)*t)):
g[3,3]:=exp(2*r*t):
g[4,4]:=-1:
```

```
metric:=create([-1,-1],eval(g)):
`tensor/simp`:=proc(x) simplify(x,trig) end:
tensorsGR(coords,metric,contra_metric,det_met,C1,C2,Rm,
Rc,R,G,C):
displayGR(cov_metric,metric);
displayGR(Ricci,Rc);
```

APPENDIX D

MAPLE PROGRAM 2

The following simple program written in MAPLE was used extensively to calculate and simplify the expressions of the various transformed metrics in both 3 and 4 dimensions. Some of the expressions run to several pages.

```
with (tensor):
# Specify coordinates
coordsx:=[x1,x2,x3,x4]; coordsy:=[y1,y2,y3,y4];
# Specify coordinate transformation
 F:=(x,y)→F(x,y): K:=(x,y)→K(x,y): H:=x→H(x):
 xtoy:=[x1=F(y1-y4,y2),x2=y2,x3=K(y1-y4,y2)+y3,x4=
 H(F(y1-y4,y2))-y4];
# Specify metric in x-coordinates
g:=array(symmetric,1..4,1..4):
 f:=x1^2+a^2*cos(x2)^2: G:=x1^2-2*m*x1+a^2:
g[1,1]:= f/G: g[1,2]:=0:g[1,3]:=0:g[1,4]:=0:
g[2,2]:=f:g[2,3]:=0:g[2,4]:=0:
g[3,3]:=((x1^2+a^2)^2-G*a^2*sin(x2)^2)*sin(x2)^2/f:
```

```
g[3,4]:=-2*a*m*x1*sin(x2)^2/f: g[4,4]:=-(1-2*m*x1/f):
gx:= create([-1,-1], eval(g)):
gx11:=g[1,1];gx12:=g[1,2];
gx13:=g[1,3]; gx14:=g[1,4]; gx22:=g[2,2];gx23:=g[2,3];
gx33:=g[3,3]; gx34:=g[3,4]; gx44:=g[4,4];
# Compute Jacobians
jacobian(coordsy, xtoy, yJx, xJy):
op (yJx): op (xJy):
# Compute metric in y-coordinates
gy:= transform(gx, xtoy, yJx, xJy):
 gy14:=simplify(gy[compts][1,4]); gy24:=simplify(gy[compts][2,4]);
 gy34:=simplify(gy[compts][3,4]); gy44:=simplify(gy[compts][4,4]);
 gy11:=simplify(gy[compts][1,1]); gy12:=simplify(gy[compts][1,2]);
 gy13:=simplify(gy[compts][1,3]); gy22:=simplify(gy[compts][2,2]);
 gy23:=simplify(gy[compts][2,3]); gy33:=simplify(gy[compts][3,3]);
```

APPENDIX E
MAPLE PROGRAM 3

The following program written in MAPLE can be used to compute the left-hand sides of Eqs. (1.41)–(1.43), given a 3-metric g and 3-vector field X. It can thus be used to verify if a given (g, X) is a solution of (1.41)–(1.43) or not.

```
with(tensor):
# Specify coordinates
coords:=[w1,w2,w3]:
# Specify metric
g:=array(symmetric,sparse,1 .. 3, 1 .. 3):
g[1,1]:=(w1^2+a^2*cos(w2)^2)/(w1^2+a^2-2*m*w1)-
2*m*w1*(w1^2+a^2)*(w1^2+a^2*cos(w2)^2-2*m*w1)/
((w1^2+a^2*cos(w2)^2)* ((w1^2+a^2-2*m*w1)^2)):
g[1,3]:=-a*sin(w2)^2*(2*m)^(3/2)*w1*((w1^3+a^2*w1))^(1/2)/
((w1^2+a^2*cos(w2)^2)*(w1^2+a^2-2*m*w1)):
g[2,2]:= w1^2+a^2*cos(w2)^2:
g[3,3]:= (w1^2+a^2-2*m*w1+2*m*w1*(w1^2+a^2)/
```

```
(w1^2+a^2*cos(w2)^2)) * (sin(w2))^2:
metric:=create([-1,-1],eval(g)):
displayGR(cov_metric,metric);
# Specify Vector field
X:=create([1],array([
 (2*m)^(1/2)*(w1^3+a^2*w1)^(1/2)/(w1^2+a^2*cos(w2)^2),
 0,2*a*m*w1/((w1^2+a^2*cos(w2)^2)*(w1^2+a^2-2*m*w1))]));
# eval(X[compts]);
# Compute its 1st and 2nd Lie-derivatives
h:=Lie_diff(metric,X,coords): N:=Lie_diff(h,X,coords):
# Simplify
# `tensor/simp`:=proc(x) simplify(x,trig) end:
# Compute Curvature, Ricci tensors etc.
tensorsGR(coords,metric,contra_metric,det_met,C1,C2,
Rm,Rc,R,G,C):
# Compute the other tensors and scalars
M:=prod(contra_metric,h,[2,2]): S:=contract(M,[1,2]):
T:=prod(M,M,[1,2],[2,1]): E:=prod(M,h,[1,2]):
F:=cov_diff(M,coords,C2):
Q:=contract(F,[1,3]): P:=cov_diff(S,coords,C2):
# Express eqn as matrices
A:=evalm(Q[compts] - P[compts]):
B:=evalm(-2*R[compts] + (1/2)*(S[compts]^2 - T[compts])):
C:=evalm(Rc[compts] - (1/2)*N[compts]
- (1/4)*S[compts]*h[compts] + (1/2)*E[compts]):
# Evaluate A, B, C
A1:=simplify(A[1]):A2:=simplify(A[2]):A3:=simplify(A[3]):
B0:=simplify(B):
```

```
C11:=simplify(C[1,1]):C22:=simplify(C[2,2]):C33:=simplify(C[3,3]):
C12:=simplify(C[1,2]):C13:=simplify(C[1,3]):C23:=simplify(C[2,3]):
A1:=A1; A2:=A2; A3:=A3; B0:=B0;
C11:=C11; C22:=C22; C33:=C33; C12:=C12; C13:=C13;
C23:=C23;
```

Here, Ai $(i = 1, 2, 3), B0, Cik$ $(i, k = 1, 2, 3)$ are the left-hand sides of Eqs. (1.41)–(1.43), respectively.

For example, the output of the above program (after about 8 to 10 minutes, depending on the machine) turns out to be:

$$A1 := 0,$$

$$A2 := 0,$$

$$A3 := 0,$$

$$B0 := 0,$$

$$C11 := 0,$$

$$C22 := 0,$$

$$C33 := 0,$$

$$C12 := 0,$$

$$C13 := 0,$$

$$C23 := 0.$$

AUTHOR INDEX

SUBJECT INDEX